POWER ELECTRONICS
Solid State Motor Control

POWER ELECTRONICS
Solid State Motor Control

Richard A. Pearman

 Reston Publishing Company, Inc., Reston, Virginia
A Prentice-Hall Company

Library of Congress Cataloging in Publication Data

Pearman, Richard A.
  Power electronics.

  Bibliography: p.
  Includes index.
  1. Power electronics. 2. Electric motors—
Automatic control. 3. Electronic control I. Title.
TK7835.P35   621.313   79-25283
ISBN 0-8359-5585-0

© 1980 by
Reston Publishing Company, Inc.
A Prentice-Hall Company
Reston, Virginia

10   9   8   7   6   5   4

PRINTED IN THE UNITED STATES OF AMERICA

Contents

harmonic reduction techniques, with the emphasis being placed on circuits and waveforms.

Chapter 7 brings out the requirements of forced commutation methods as applied to dc-dc control. Chapter 8 is devoted to firing circuit requirements and methods.

Chapter 9 concentrates on the application of power electronics to ac and dc motor applications, and introduces new methods, such as dc brushless motors and the microprocessor.

The information presented in this book has been acquired over a number of years of teaching, and studying industrial applications in local industry.

The author wishes to acknowledge his indebtedness to John Vidoli for his careful and thorough review of the material, and my wife and daughter for their patience in the many hours of typing and checking that they have contributed.

RICHARD A. PEARMAN

Preface

Since 1957 with the introduction of the thyristor, combined with the subsequent development of integrated circuits, digital logic techniques, and microprocessors, the application of solid-state electronic devices has made considerable inroads in the control of ac and dc rotating equipment and industrial processes. The combined result has been the creation of a new field in the electrical industry, namely, power electronics. This in turn has spawned a series of texts, handbooks, and technical papers, which usually are very specialized or give a very basic treatment of many topics.

The purpose of this book is to provide a good understanding of power electronics to meet the needs of two-year technician and three-year technologist level programs at community colleges, and in-house industrial training programs.

Chapter 1 is intended to give an overview of the role of the thyristor in power electronic applications. Chapter 2 establishes the theory, characteristics, and protection of the thyristor. Chapter 3 deals with ac and dc static control applications, and this is followed by ac phase control in Chapter 4.

Chapter 5 deals with controlled rectification, with the emphasis being placed on circuits and waveform analysis of the operation of one- and two-quadrant converters.

Chapter 6 studies static frequency conversion by means of the dc link converter and the cycloconverter, emphasizing frequency, and voltage control and

POWER ELECTRONICS
Solid State Motor Control

1   A Review of Thyristor Applications

1-1   INTRODUCTION

The requirement for reliable and versatile control of electric motor drive systems has existed for many years. Initially this need was met in the late nineteenth century with the development of the Ward-Leonard system for dc motor drives. The impetus for the development of the Ward-Leonard concept grew with the introduction of industrial automation and remote controlled gun drive systems during World War II.

The initial Ward-Leonard concept was continuously modified and improved with the introduction of the thyratron, the mercury pool rectifier, the ignitron, and the excitron during the 1940s. In the early 1950s with the introduction of the transistor and the magnetic amplifier, improved closed-loop controls were applied to dc motor systems. In the latter part of the 1950s and during the 1960s the thyristor began to replace the gaseous-discharge tubes as the source of armature voltage control for dc motors. In the latter part of the 1960s and into the 1970s, the control elements were replaced by the versatile operational amplifier, logic elements, and now the microprocessor.

The modern version of the Ward-Leonard concept is being used for speed and position control of dc adjustable speed drives from fractional horsepower portable tools to sizes in excess of 10,000 hp (7460 kW), in such applications as machine tools; the steel, textile, and paper industries; mine hoists; electric traction; cranes and elevators. The limitations are imposed by the speed and horsepower capabilities of the dc motor, not by the solid-state devices.

Compared to the polyphase ac squirrel cage induction motor, the initial cost of a dc motor of comparable size is about three to four times as much, but the most important economic factor is the high maintenance cost of the dc motor as compared to the ac motor over the projected life span, which has led to a great deal of research in the development of ac variable-speed drive systems.

The major problem with the application of the polyphase induction motor was to obtain continuously variable-speed control over a wide range. Traditional methods such as pole changing and pulse amplitude modulation gave limited speed control but at several discrete speeds.

In the 1930s and 1940s the concept of supplying variable frequency voltages was developed in Europe through the use of variable frequency inverters and cycloconverters with gaseous-tube devices, such as the mercury pool rectifier and the ignitron. In the 1960s, with the availability of the silicon controlled rectifier (SCR), great advances were made in continuously variable ac adjustable speed drives, resulting in higher efficiencies, greater system reliability, and reduced initial and maintenance costs as compared to dc adjustable drive systems. AC variable-speed drives are now available from fractional horsepower sizes to 20,000 hp (14920 kW).

1-2   THE SILICON CONTROLLED RECTIFIER (SCR)

The most extensively used solid-state device in the power electronics field is the thyristor, which can be considered as a power amplifier with a power handling capability varying from 40 W (1.6 A, 25 V) to 4 MW (2500 A, 1600 V) for converter and inverter applications. Some thyristors having ratings of 400 A at 4000 V and 10,000 V have also been developed for high voltage direct current transmission systems.

The thyristor family, of which the SCR is one member, consists of some 25 different four-layer silicon devices. The SCR is a fast switching, unidirectional power switch presenting a high impedance in the OFF state and a very low impedance in the ON state. The device is switched on by applying a low voltage, low current positive signal to the gate terminal when the anode-cathode is forward-biased.

1-3   STATIC SWITCHING

The SCR and the bidirectional triode thyristor (TRIAC) are bistable devices that are extensively used in signal and power switching. These thyristor devices are now being applied in areas that were the traditional stronghold of mechanical and electromagnetic switches.

Static switching applications break down into two main classifications: ac line control and dc line control.

1-3-1   AC Line Control

The simplest method of controlling ac power to a load is to use two SCRs in inverse parallel, as in Fig. 1-1(a), or a TRIAC, as in Fig. 1-1(b).

The applied ac voltage V_{AB} is shown in Fig. 1-1(c), with point A being positive for one-half cycle and point B being positive for the next half-cycle, with SCR1 being forward-biased for the first half-cycle and SCR2 being reverse-biased, and for the next half-cycle the situation being reversed.

As long as the SCRs or TRIACs are blocked, the voltage at the load is zero. If, as in Fig. 1-1(d), SCR1 is triggered at t_1, the voltage at the load will be equal to V_{AB} minus the very small drop across the SCR; at t_2 with a pure resistive load SCR1 will commutate off, and the load voltage will be zero until SCR2 is triggered at t_3. By controlling the point at which the SCRs are gated the mean ac voltage supplied to the load can be controlled.

In the case of an inductive load with SCR1 gated on at t_1, at t_2, SCR1 will remain conducting with a forward current until the stored magnetic energy has been dissipated, at which point SCR1 will be reverse-biased and turn off. During the interval $t_2 - t_3$ current is being fed back to the power supply from

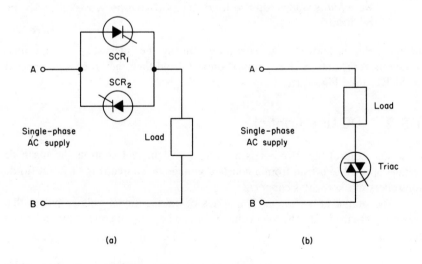

(a)   (b)

FIG. 1-1 AC line control. (a) inverse parallel connected SCRs.

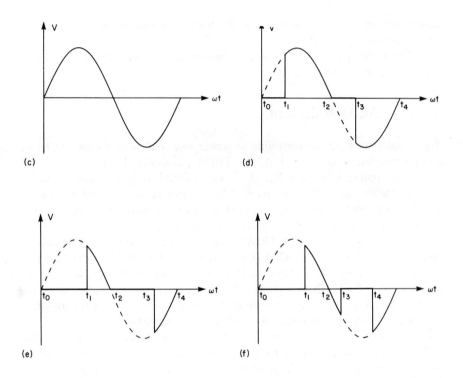

FIG. 1-1 *(cont.)* (b) TRIAC; (c) line voltage waveform; (d) and (e) load voltage
waveforms with a resistive load; (f) load voltage waveform with an
R-L load.

the load; this is known as *regeneration*. During the interval $t_3 - t_4$ the load
voltage is zero, until SCR2 is turned on at t_4, and then the process is repeated
for SCR2. [See Fig. 1-1(f).]

1-3-2   DC Line Control

In dc line control the thyristor is used as a switch, and controls the mean dc
voltage applied to a load from a constant voltage dc source, Fig. 1-2. DC to dc
controllers are known as *choppers*.

The output of a chopper is a series of unidirectional voltage pulses that
are applied to the load. The load voltage V_{do} can be varied in one of the following
ways:

1. t_{ON} variable, t_{OFF} variable, and the periodic time T constant, pulse
width modulation.

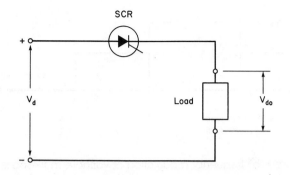

FIG. 1-2 DC-DC (chopper) converter, basic principle.

2. t_{ON} constant, t_{OFF} variable, and the periodic time T is varied, frequency modulation or variable mark-space.

3. A combination of (1) and (2).

These principles are illustrated in Fig. 1-3. The mean load voltage over a cycle is given by

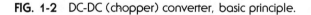

$$V_{do} = \frac{t_{ON}}{t_{ON} + t_{OFF}} \cdot V_d = \frac{t_{ON}}{T} \cdot V_d \tag{1-1}$$

where T = the periodic time

The load voltage $V_{do} = 0$ when $t_{ON} = 0$ and is equal to the source voltage when $t_{OFF} = 0$.

Choppers provide an excellent means of controlling dc traction motors, since they eliminate the energy losses in starting and control resistances. As we

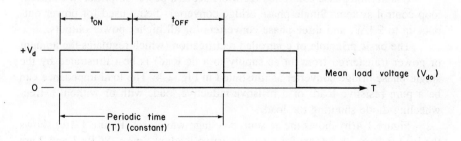

FIG. 1-3 DC-DC (chopper) converter modes of operation. (a) pulse width modulation.

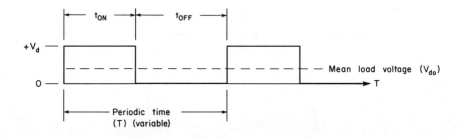

FIG. 1-3 *(cont.)* (b) frequency modulation or variable mark-space.

illustrate the principle of chopper control, remember that since the SCR is supplied from a dc source, means must be provided by a forced commutation circuit to turn off the SCR at the end of the conduction period, t_{ON}.

1-4   CONTROLLED RECTIFICATION

By far the largest application of thyristors is in controlled rectification. The mean dc voltage applied to the load is supplied by phase controlled thyristors. Some applications of phase controlled rectifiers are:

1. Variable armature voltage speed control of dc motors in the steel and paper industries.
2. Variable dc sources for dc-link converters used in variable frequency control of induction motors.
3. Variable speed control of portable power tools.

Usually the phase controlled rectifiers act as a power amplifier in a closed-loop control system. Single-phase bridge converters being used for power outputs up to 2 kW, and three-phase converters for all higher power outputs.

The basic principle of controlled rectification (which regulates the amount of power transferred from an ac supply to a dc load) is best illustrated by the single-phase bridge converter as illustrated in Fig. 1-4. The load impedance can be a pure resistive load, or a resistive-inductive load, with or without a free-wheeling diode shunting the load.

Figure 1-4(b) shows the ac source voltage waveform. Figure 1-4(c) shows the load voltage waveform for a pure resistive load; at time t_1 SCRs 1 and 2 are turned on simultaneously, and the voltage supplied to the load is the ac source voltage less the device voltage drops. At time t_2, SCRs 1 and 2 are reverse-biased and turn off, and load current ceases. At time t_3, SCRs 3 and 4 are turned

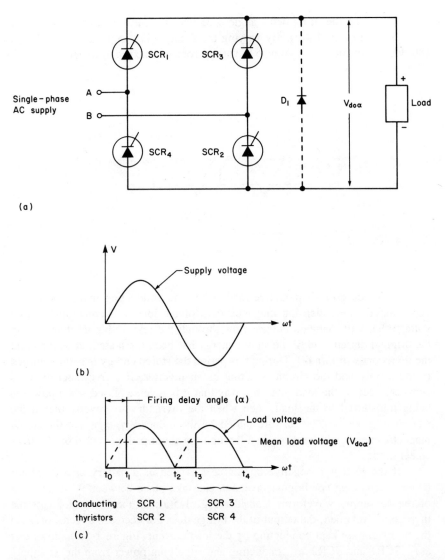

FIG. 1-4 Controlled rectification. (a) single-phase bridge converter; (b) supply voltage waveform (c) and (d) load voltage waveforms with variable firing delay.

on, and even though the ac source voltage has reversed, the voltage applied to the load has the same polarity as when SCRs 1 and 2 were conducting—that is, the load voltage is unidirectional, and has a mean amplitude $V_{do\alpha}$. The interval t_0 to t_1 is known as the firing delay angle α.

Increasing the firing delay angle α reduces the mean output voltage $V_{do\alpha}$, as is shown in Fig. 1-4(d). By varying the firing delay angle α from 0 deg to 180 deg the mean output voltage is varied from its maximum to zero.

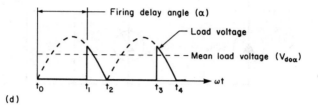

(d)

FIG. 1-4 *(cont.)*

In the case of an inductive load, when the conducting thyristors are reverse-biased, the collapsing magnetic field of the load will create an induced voltage that will maintain current flow through the reverse-biased thyristors in the original direction until the stored energy has been dissipated, at which point the thyristors will turn off. During this period the stored energy is being returned to the supply and the circuit is acting as an inverter. If a freewheel diode is connected across the load in such a way that it is reverse-biased when power is being transferred to the load, then when the thyristors are reverse-biased the freewheel diode will be forward-biased by the induced voltage and the stored magnetic energy will be dissipated through the shunt path provided by the freewheel diode.

If the load is a negative dc source E_c, such as a battery or a dc motor under overhauling conditions, inversion can take place between 90 to 180 deg of the ac supply waveform. Consider Fig. 1-5(a); as the firing point of the thyristors is retarded, the output of the bridge does not decrease to zero, because the thyristors are kept conducting by the load current. Figure 1-5(c) shows that at t_1, SCR1 and SCR2 are conducting and supplying power from the ac supply to the load; at t_2, SCR1 and SCR2 remain conducting, carrying current supplied by the load voltage; at t_3, SCR3 and SCR4 are turned on and SCR1 and SCR2 are reverse-biased and turn off. For $0° < \alpha < 90°$, the mean output voltage $V_{do\alpha}$ is positive; that is, there is a net transfer of power from the ac supply to the load. For $90° < \alpha < 180°$, the mean output voltage $V_{do\alpha}$ is negative, and there is a net transfer of power from the load dc source to the ac supply [see Fig. 1-5(d)].

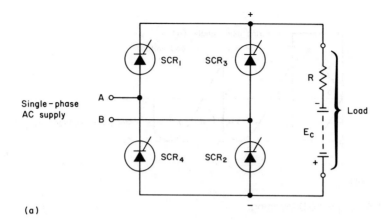

(a)

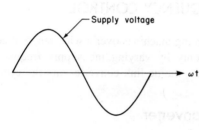

(b)

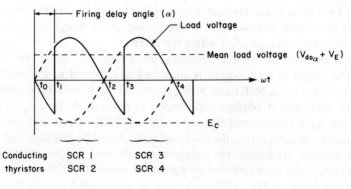

FIG. 1-5 Controlled rectifier-inversion mode. (a) single-phase bridge converter;
(b) supply voltage waveform; (c) rectifying.

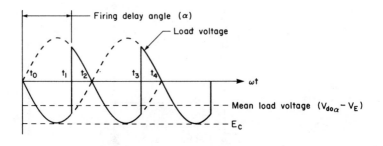

(d)

FIG. 1-5 *(cont.)* (d) inverting.

1-5   VARIABLE FREQUENCY CONTROL

Speed control of polyphase rotating machines over a wide range of continuously variable speeds is attainable only by varying the supply frequency. This is achieved by two major methods, the dc link converter and the cycloconverter.

1-5-1   The DC Link Converter

The dc link converter converts power from the ac supply to a rectified dc, then inverts the dc to ac at the desired frequency. Usually the rectification is accomplished by a three-phase thyristor bridge, phase control being used to obtain a variable dc output; alternatively, a three-phase diode bridge is used and the output ac voltage is controlled within the inverter.

The principle of variable frequency speed control is illustrated in Fig. 1-6(a). The output of a dc source is smoothed and applied to the single-phase inverter. If at time t_0, SCR1 and SCR2 are turned on, V_d will be applied across the load with point A positive with respect to point B. At time t_1, SCR1 and SCR2 are force commutated off and SCR3 and SCR4 are turned on; then V_d will again be applied across the load, but point A will be negative and point B will be positive. As a result the voltage across the load will be alternating, and the frequency will be determined by the rate at which the SCR pairs are turned on and off [refer to Fig. 1-6(b)]. It must be appreciated that in this type of inverter forced commutation techniques must be used, since the inverter is supplied from a dc source, whereas the inversion process, described in Fig. 1-5, occurs naturally as the ac source voltage reverse-biases the thyristors.

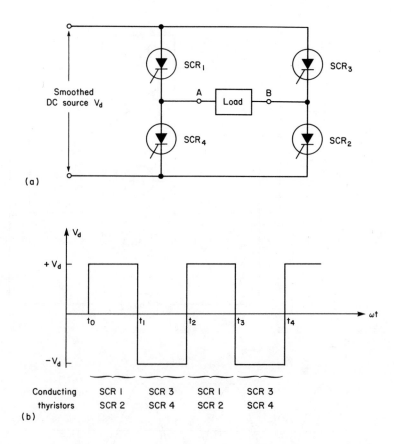

FIG. 1-6 Single-phase inverter. (a) basic circuit; (b) load voltage waveform.

1-5-2   The Cycloconverter

The cycloconverter converts ac at one frequency directly to a lower frequency without conversion to dc as in the inverter. Figure 1-7 shows the basic principle of a cycloconverter.

The cycloconverter depends upon the cyclic variation of the firing point of the thyristors in the positive and negative groups. The cyclic variation of the firing point is usually obtained from a low frequency sinusoidal signal being applied to the gate firing circuitry of the cycloconverter. The variation of the mean output voltage has the same frequency as the signal applied to the gating circuits.

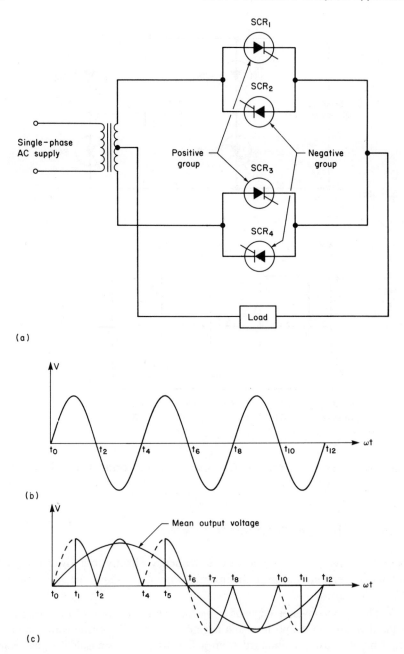

(a)

(b)

(c)

FIG. 1-7 Single-phase cycloconverter, (a) basic circuit, (b) supply voltage waveform, (c) load voltage waveform.

The cycloconverter output is usually limited to frequencies below one-third of the supply frequency. Another disadvantage is that a minimum of 18 SCRs is required for unidirectional operation of a three-phase ac drive, and 36 SCRs are required for bidirectional operation. However, the cycloconverter is finding a number of applications in low-speed motor control, and currently is being used to control synchronous motors in the 8000-hp (6000 kW) range for the direct drive of cement kilns.

1-6   SUMMARY

The purpose of this chapter has been to introduce the concepts of thyristor applications to the control of ac and dc power as well as the four basic power conversion systems, namely dc to dc, ac to dc, dc to ac, and ac to ac. In subsequent chapters each of these topics will be dealt with in greater detail.

2 The Thyristor

2-1 INTRODUCTION

Thyristor is a generic term for any semiconductor device that exhibits the regenerative switching characteristic of a four-layer or p-n-p-n arrangement. A number of devices such as the DIAC (bidirectional diode thyristor), the SUS (silicon unilateral switch), the SBS (silicon bilateral switch), the SCR (silicon controlled rectifier, reverse-blocking triode thyristor), and the TRIAC (bidirectional triode thyristor) are some of the many members of the thyristor family. The most important member of the thyristor family is the silicon controlled rectifier (SCR), which was developed by General Electric in 1957.

The SCR is a three-terminal, three-junction, four-layer (p-n-p-n) semiconductor device. Because of the three p-n junctions in the device it can be seen that there is at least one reverse-biased p-n junction opposing conduction in either direction, thus giving the SCR blocking capabilities in either direction. Functionally the SCR has replaced the grid-controlled mercury arc rectifier and the thyratron because of its small size and weight, maintenance-free operation, the capability of being operated in any position, reliability, and the low cost per ampere. Thyristors are commercially available in current ratings ranging from milliamperes to in excess of 2500 A, with voltage ratings ranging from a low of 15 V to as high as 2500 V. Hitachi Research Laboratory has developed 4000-V, 400-A and 10,000-V, 400-A thyristors for use in high-voltage dc transmission. As it can be seen, the thyristor can control megawatts of power.

Thyristors are now available in a number of different configurations (see Fig. 2-1), from the chip form to the disc type or hockey-puck form with or without integral heat sinks depending on the heat load that must be dissipated.

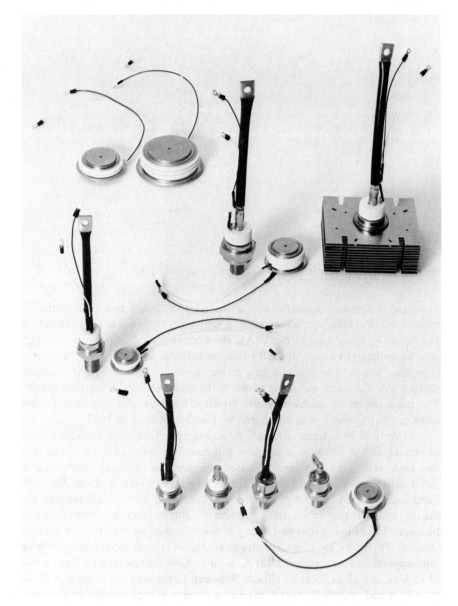

FIG. 2-1 Typical SCR configurations. (*Courtesy Westinghouse Corp.*)

The TRIAC is an extensively used device in ac line control. Most TRIACs are available in ratings of less than 40 A and at voltages up to 600 V; The International Rectifier Corp. produces 60-, 100-, and 200-A devices rated as high as 1000 V.

2-2   THE SILICON CONTROLLED RECTIFIER (SCR)

The most extensively used member of the thyristor family is the reverse-blocking triode thyristor or silicon controlled rectifier, the SCR. It is a four-layer, three-terminal, three-junction device with two power terminals and one control terminal. The layers are formed by diffusion or alloying. The major difference as compared to the diode is that forward conduction normally does not take place until a low-power firing signal is applied to the control terminal or gate. The reverse characteristic is similar to that of the silicon diode.

Essentially the device is a switch, and presents a high forward impedance until both the anode and gate terminals are positive with respect to the cathode, but when triggered by a positive signal being applied to the gate, it goes into a low impedance state. It will remain in conduction until the anode-cathode current I_A is reduced below a level called the holding current I_H.

It is also possible to turn the device on by increasing the forward voltage until junction breakdown occurs; this condition is called forward breakover, V_{FBO}.

2-2-1   The Two-transistor Analogy of the SCR

The most common method of explaining the operation of the SCR is in terms of the two-transistor analogy as shown in Fig. 2-2. The basic construction, shown in Fig. 2-2(a), is a four-layer diode. The center section can be divided as shown in Fig. 2-2(b) and (c). The SCR can then be considered as two separate complementary transistors, one a p-n-p transistor and the other an n-p-n transistor, Q_1 and Q_2 respectively. If the gate signal is zero and the anode is positive or negative with respect to the cathode, one of the p-n junctions in each transistor is reverse-biased. When the SCR is forward-biased, junctions J_1 and J_3 are forward-biased and J_2 is reverse-biased; conversely, when the thyristor is reverse-biased, junctions J_1 and J_3 are reverse-biased and J_2 is forward-biased, and only a small leakage current flows.

The application of a positive gate signal when the thyristor is reverse-biased causes a reverse-anode leakage current to flow approximately equal to the positive gate current, and junction overheating can result in a thermal runaway.

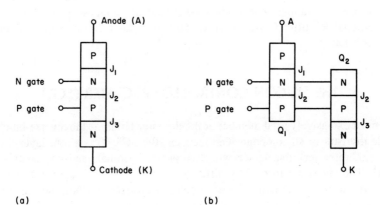

(a)   (b)

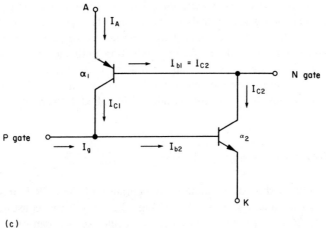

(c)

FIG. 2-2   The two-transistor equivalent circuit of the thyristor. (a) basic structure;
(b) as two complementary transistors; (c) two transistor equivalent circuit.

Using the two-transistor analogy, as shown in Fig. 2-2(b) and (c), J_1 and
J_3 are slightly forward-biased, where J_1 is the emitter-base junction of the p-n-
p transistor and J_3 is the emitter-base junction of the n-p-n transistor. J_2 is the
collector-base junction of both Q_1 and Q_2 and is reverse-biased. Since the emitter-base junctions are only slightly forward-biased, there will be little current
flow.

Consider the p-n-p transistor Q_1; its emitter current is I_A, the anode current, and the base current is

$$I_{b1} = (1 - \alpha_1)I_A - I_{CBO1} \tag{2-1}$$

where α_1 is the current gain and I_{CBO1} is the leakage current for Q_1.

Now consider the n-p-n transistor Q_2:

$$I_{C2} = \alpha_2 I_K + I_{CBO2} \tag{2-2}$$

where α_2 and I_{CBO2} are current gain and leakage current respectively for Q_2. I_K, the cathode current, is equal to the emitter current I_E of Q_2.

Since the base current of Q_1, I_{b1}, and the collector current I_{C2} of Q_2 are the same, then

$$(1 - \alpha_1)I_A - I_{CBO1} = \alpha_2 I_K + I_{CBO2} \tag{2-3}$$

and since $I_A = I_K$, then

$$I_A = I_K = \frac{I_{CBO1} + I_{CBO2}}{1 - (\alpha_1 + \alpha_2)} \tag{2-4}$$

Equation (2-4) forms the basis of explanation for all p-n-p-n devices. When both Q_1 and Q_2 have a very small forward bias of the emitter-base junction, the value of α is $<<1$, and $(\alpha_1 + \alpha_2)$ is small, and I_A will be small. The sum of $(\alpha_1 + \alpha_2)$ can be made momentarily to approach 1 by injecting a short duration positive current I_g at the P gate, which is the base of Q_2. This causes current to flow in Q_2 and because the collector is positive, collector current will flow in Q_2; this is also the base current of Q_1, and as a result Q_1 will be switched on. At this point, each transistor supplies the base current for the other transistor and the action is regenerative. Removal of the gate signal will not result in the thyristor turning off as long as there is sufficient forward anode-to-cathode voltage to maintain a holding current I_H. From the above it can be seen that a small-amplitude positive pulse of a few microseconds duration applied when the anode-cathode is forward-biased will ensure turn-on of the thyristor. However, once conduction has been initiated, the gate signal serves no useful purpose and may be removed.

2-2-2   Initiation of Thyristor Turn-on

Basically a thyristor is turned on by causing an increase in the emitter current. This action can result from any of the following:

1. Gate current I_g: The application of a positive pulse of sufficient amplitude and duration to the P gate of Fig. 2-2.

2. Overvoltage: An increase in the forward anode-cathode voltage above the forward breakover voltage, V_{FBO}, will cause a sufficient increase in the leakage current to initiate regenerative turn-on.

3. dv/dt: The p-n junctions are effectively capacitive because of the depletion layer during blocking. As a result, whenever there is a rapid rate of change of the anode-cathode voltage (dv/dt), the charging current $i = C dv/dt$ may be of sufficient magnitude when added to the leakage current to initiate turn-on.

4. Thermal: There is an increase in the number of electron-hole pairs as the temperature of the device increases, which causes an increase in ($\alpha_1 + \alpha_2$) resulting in turn-on being initiated at lower forward anode-cathode potentials.

5. Light or radiation: Photons, gamma rays, neutrons, protons, electrons, and hard and soft X-rays when permitted to strike an unshielded thyristor will cause an increase in the electron-hole pairs, thus initiating turn-on. In some devices, such as the light-activated SCR (LASCR), a window is provided in the shielding to permit the thyristor to be turned on by allowing light to strike the silicon wafer.

2-3   SCR RATINGS AND CHARACTERISTICS

The basic characteristics of the SCR are shown in Fig. 2-3, which is a plot of anode current (I_A) vs. anode-cathode voltage (V_{AK}).

Note: the technical and graphical terminology used in the text is described in Table 2-1, the Westinghouse terminology being used with the JEDEC equivalent being shown where applicable.

2-3-1   Voltage Parameters

With a zero gate current, the device acts as an open circuit; that is, it is in the high impedance state until a sufficiently high anode-cathode voltage, known as the forward breakover voltage V_{FBO} ($V_{(BO)}$) is reached. At this point the leakage currents increase sufficiently so that the device turns on. The forward breakover voltage test and all the subsequent voltage rating tests are carried out with the device operating at rated junction temperature. It should be noted that the SCR is not designed to be turned on by the forward breakover voltage, V_{FBO}, ($V_{(BO)}$) and if it is turned on by this method the SCR can be damaged.

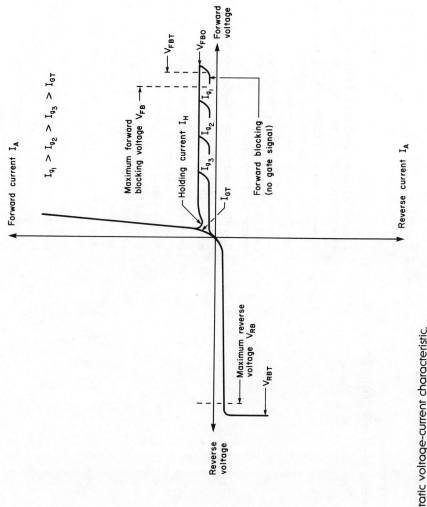

FIG. 2-3 Static voltage-current characteristic.

21

TABLE 2-1 (*Courtesy of Westinghouse Corp.*)

3.1.1 Voltage Terminology

JEDEC	Westinghouse Term	MIL-S-19500	Other Use	
V_{DRM}	V_{FB}	V_{FBOM}	PFV	**Repetitive Forward Blocking Voltage** – The maximum instantaneous value of forward voltage which occurs across the thyristor, including all repetitive voltages.
V_{DSM}	V_{FBT}	V_{ROM}	V_{ROM}	**Non-repetitive Peak Off-State Voltage** (Forward Transient) The maximum instantaneous value of any non-repetitive transient off-state voltage which occurs across the thyristor.
$V_{(BO)}$	V_{FBO}	$V_{(BO)FBO}$	$V_{(BR)F}$	**Breakover Voltage** – The minimum value of voltage that will cause the device to conduct. This voltage is further amended to identify specific gate and dv/dt conditions where applicable.
V_{RRM}	V_{RB}	$V_{RM(rep)}$	$V_{ROM(rep)}$	**Repetitive Peak Reverse Voltage** (of a Reverse Blocking Thyristor) The maximum instantaneous value of the reverse voltage which occurs across the thyristor, including all repetitive transient voltages.
V_{RSM}	V_{RBT}	$V_{RM(non-rep)}$	V_{RSM}	**Non-repetitive Peak Reverse Voltage** (of a Reverse Blocking Thyristor) The maximum instantaneous value of any non-repetitive transient reverse voltage which occurs across a thyristor.
V_{TM}	V_F	V_{FM}		**Forward Voltage Drop** – The instantaneous voltage drop between the anode and cathode of the SCR. This voltage is measured for load currents and times such that local turn-on transients are not included.
$V_{T(av)}$	$V_{F(ave)}$	$V_{F(av)}$	$V_{F(av)}$	**Average Voltage Drop** – The average value of on-state voltage of an SCR, with stated average current and test conditions.
	$V_{(DYN)}$		V_{TO}	**Dynamic Voltage Drop** – The transient forward on-state voltage of the SCR as it is triggered into conduction for given load and time conditions.
	dv/dt		dv/dt	**Critical Rate of Rise of Off-State Voltage** – The minimum value of the rate of rise of forward voltage which will cause switching from the off-state to the on-state.

dv/dt
(reapplied)

Reapplied Rate of Rise of Off-State Voltage – The minimum value of the rate of rise of forward voltage which will cause a commutation failure under specific circuit commutated turn-off conditions.

3.1.2 Current Terminology

JEDEC	Westinghouse Term	MIL-S-19500	Other Use	
I_{DRM}	I_{FB}	i_{FBOM}	I_{DM}	**Forward Blocking Current** – The maximum current that occurs with repetitive V_{DRM} testing.
I_{DSM}	I_{FBT}	i_F		**Forward Blocking Transient Current** – The maximum value of current that is experienced in V_{DSM} testing.
I_{RRM}	I_{RB}	i_{RBOM}	I_{RM}	**Reverse Blocking Current** – The maximum current that occurs with V_{RRM} testing.
I_{RSM}	I_{RBT}		i_R	**Reverse Blocking Transient Current** – The maximum value of current that is experienced with V_{RSM} testing.
$I_{T(RMS)}$	I_{RMS}		I_F	**Conducting RMS Current** – Maximum value of RMS current that the SCR may conduct.
$I_{T(av)}$	I_{ave}	I_o	$I_{F(ave)}$	**Conducting Average Current** – The maximum continuous average value of current that the SCR may conduct for various conduction angles.
I_{TM}	I_F	i_{FM}		**Maximum Forward On-State Current** – The SCR current that is flowing at V_{TM} testing.
I_H	I_H	I_{HOO}	i_H	**Holding Current** – That value of forward anode current which allows the device to remain in conduction. Below this value the device will return to a forward blocking state at the prescribed gate conditions.
I_L	I_L			**Latching Current** – The minimum value of anode current required to keep the device in conduction after the removal of a gate pulse.

23

TABLE 2-1

JEDEC	Westinghouse Term	MIL-S-19500	Other Use	
	$I_{R(REC)}$		I_{RQM}	**Reverse Recovery Current** – The peak value of reverse current that occurs during the junction recovery interval.
I_{TSM}	I_{FM}	$i_{FM(surge)}$		**Surge Current** – The peak value of a single half cycle (i.e. 180° of a sine wave) current impulse at 60 Hz. This rating is non-repetitive and may occur 100 times within the life of the device. NOTE: Following this current surge the V_{DRM} rating is not guaranteed.
di/dt	di/dt		di/dt	**Critical Rate of Rise of On-State Current** – The maximum value of the rate of rise of on-state current which an SCR can withstand without being harmed.
di_R/dt	di_R/dt		di_R/dt	**Rate of Decay of On-State Current** – The maximum value of the rate of decay of on-state current which an SCR may experience in a particular test and rating.

3.1.3 Gate Parameters

JEDEC	Westinghouse Term	MIL-S-19500	Other Use	
I_{GT}	I_{GT}	I_{GT}	I_{GT}	**Gate Trigger Current** – The minimum DC gate current required to switch an SCR from off-state to on-state.
I_{GD}	I_{GNT}	I_{GNT}	I_{GNT}	**Gate Non-Trigger Current** – The maximum DC gate current which will not cause the SCR to switch from the off-state to the on-state.
i_{GFM}	i_{GFM}	i_{GKM}	$+I_{GM}$	**Peak Forward Gate Current** – The maximum value of gate current that may be used to bring the SCR into conduction.
i_{GR}	i_{GR}	i_{kGM}	I_{GRM}	**Reverse Gate Current** – The maximum value of gate current that the device cathode-gate junction may carry in the reverse direction without damage.
V_{GT}	V_{GT}	V_{GT}	V_{GT}	**Gate Trigger Voltage** – The gate DC voltage required to produce the gate trigger current.
V_{GD}	V_{GNT}	V_{GNT}		**Gate Non-Trigger Voltage** – The maximum DC gate voltage which will not cause the SCR to switch from the off-state to the on-state.

JEDEC		Westinghouse
V_{GRM}	V_{KGM}	V_{GRM}

Peak Reverse Gate Voltage — The maximum value of reverse bias that the gate cathode junction will support without damage for all bias conditions.

P_{GM}	P_{GM}	P_{GM}

Peak Gate Power – The maximum instantaneous product of gate current and gate voltage allowed to exist during any or all forward bias conditions.

$P_{G(AV)}$	$P_{G(AVE)}$	

Average Gate Power – The maximum allowable value of gate power dissipation that the device gate junction can dissipate. This value is obtained by averaging the gate power over a full cycle.

P_{GMR}		

Peak Reverse Gate Power – The maximum instantaneous product of reverse gate current and reverse gate voltage allowed to exist during any or all bias conditions.

$P_{GR(AVE)}$		

Average Reverse Gate Power – The maximum allowable value of gate power dissipation that the device gate junction can safely dissipate. This value is obtained by averaging the reverse gate power over a full cycle.

3.1.4 Power Conditioning Terminology

JEDEC	Westinghouse Term	MIL-S-19500	Other Use	
I^2t	I^2t		I^2t	**I Squared t** – This term describes the maximum, forward, non-repetitive over-current capability of the SCR. It is normally used to describe device capabilities for ½ cycle of 60 Hz current impulse.
T_J	T_j	T_j		**Operating Junction Temperature** – The virtual junction temperature of the device as a result of ambient and load current conditions. Unless specified, this parameter has a maximum value of 125°C.
T_{stg}	T_{stg}	T_{stg}		**Storage Temperature** – The temperature at which the device may be stored without harm.
T_C	T_C	T_C		**Case Temperature** – The temperature of the device case under specified load conditions.

TABLE 2-1

JEDEC	Westinghouse Term	MIL-S-19500	Other Use	
T_A	T_A $P_{AVE(MAX)}$	T_A	T_A	**Ambient Temperature** **Max. Device Dissipation** – Average power dissipation in the SCR with maximum forward drop and specified conduction intervals.
$R\theta$	θ		θ	**Thermal Resistance** – The temperature difference between two specified points or regions divided by the power dissipation under conditions of thermal equilibrium.
$R\theta JC$	θ_{jc}		θ_{jc}	**Thermal Resistance** – Junction to case.
$R\theta CS$	θ_{cs}		θ_{cs}	**Thermal Resistance** – Case to sink.
$R\theta JA$	θ_{ja}		θ_{ja}	**Thermal Resistance** – Junction to ambient.
$Z\theta(t)$	$\theta_{(t)}$		$\theta_{(t)}$	**Transient Thermal Impedance** – The change in temperature between two specified points or regions at the end of a time interval divided by the step function change in power dissipation at the beginning of the same time interval causing the change of temperature.
$Z\theta JC(t)$	$\theta_{jc(t)}$		$\theta_{jc(t)}$	**Transient Thermal Impedance** – Junction to case.
$Z\theta JA(t)$	$\theta_{ja(t)}$		$\theta_{ja(t)}$	**Transient Thermal Impedance** – Junction to ambient.

3.1.5 Switching Time Terminology

JEDEC	Westinghouse Term	MIL-S-19500	Other Use	
t_q	t_{off}	t_{off}	t_q	**Circuit-Commutated Turn-Off Time** – The time interval between the instant when the SCR current has decreased to zero after external switching of the SCR voltage circuit, and the instant when the thyristor is capable of supporting a specified voltage wave form without turning on.

t_{on}	**Gate Controlled Turn-On Time** – The time interval between a specified point at the beginning of the gate pulse and the instant when the device voltage (current) has dropped (risen) to a specified low (high) value during switching of an SCR from off-state to the on-state by a gate pulse.
t_d	**Gate Controlled Delay Time** – The time interval between a specified point at the beginning of the gate pulse and the instant when the device voltage (current) has dropped (risen) to a specified value near its initial value during switching of an SCR from the off-state to the on-state by a gate pulse.
t_r	**Gate Controlled Rise Time** – The time interval between the instants at which the device voltage (current) has dropped (risen) from a specified value near its initial value to a specified low (high) value during switching of an SCR from the off-state to the on-state by a gate pulse.
	NOTE: This time interval will be equal to the rise time of the on-state current for pure resistive load only.
t_f	**Fall Time** – The fall time is the time interval between the time the current decreases from 90% of its initial value to 10% of its initial value during switching from the on-stage to the off-state under specified conditions.
t_{gr}	**Gate Recovery Time** – The gate recovery time is the minimum time interval between the reverse recovery point and the time the device voltage crosses zero at a specified rate of voltage rise without conducting.
t_{rr}	**Reverse Recovery Time** – The time required for the device current or voltage to recover to a specified value after instantaneous switching from an on-state to a reverse voltage or current.
t_s	**Storage Time** – The storage time is the time interval between the time the gate current turn-off pulse reaches 50% of its final value and the time when the resulting forward anode current decreases to 90% of its initial value.

3.1.6 Gate Turn-Off Terminology

V_{GQM}	**Gate Turn-off Voltage** – Maximum negative gate voltage required to switch the anode circuit from the on-state to the off-state.
I_{GQM}	**Gate Turn-Off Current** – Maximum negative gate current required to switch the anode circuit from the on-state to the off-state.
	Turn-Off Current Gain – Ratio of the anode current turned off to the negative gate current required to turn off.

Normally the SCR should not be subjected to an instantaneous repetitive forward blocking voltage greater than V_{FB}, (V_{DRM}); however, for short periods of time, it will maintain its high impedance state when subjected to forward transient voltages, V_{FBT} (V_{DSM}), in excess of V_{FB} (V_{DRM}), provided the time specified in the manufacturer's data sheet is not exceeded (see Fig. 2-4).

When the anode-cathode voltage is reversed, the SCR operates as a conventional silicon diode, and there are two voltage ratings that must be observed. First, the reverse voltage, V_{RB} (V_{RRM}), which is the maximum instantaneous repetitive reverse voltage, including repetitive transient voltages that may be applied without initiating avalanche breakdown. Secondly, the device is rated for the non-repetitive peak reverse transient voltage V_{RBT} (V_{RSM}), but once again this must not be applied for a longer time than specified in the data sheets since the device is operating in the avalanche mode and excessive heating of the device will occur.

2-3-2   Gate Parameters

The gate parameters can be classified in terms of current and voltage. I_{GT} is the minimum dc gate current required to cause an SCR to switch from the high impedance state to conduction. I_{GNT} is the maximum dc gate current that will not cause the SCR to turn on, i_{GFM} is the maximum gate current that is permissible to be applied to achieve turn-on, and i_{GR} is the maximum reverse gate current that is permitted to be applied to the gate-cathode junction without causing damage. The gate current parameters are based on the device being operated at the specified junction temperatures and normal ambient temperatures.

The corresponding gate voltage characteristics are as follows: the gate trigger voltage V_{GT} which is the dc voltage necessary to produce I_{GT}, V_{GNT} (V_{GD}), which is the maximum dc gate voltage that will not result in the SCR being turned on, and finally, V_{GRM}, which is the maximum value of negative dc voltage that may be applied without damaging the gate-cathode junction.

In addition, the power dissipation capabilities of the gate junction must be defined in order to prevent overheating. The applicable gate power parameters are the peak gate power P_{GM}, which is the maximum instantaneous product of the gate current and gate voltage that is permitted to exist during forward-bias conditions. The average gate power, $P_{G(AVE)}$, $(P_{G(AV)})$, is the maximum gate power dissipation that is permitted at the gate junction over a full cycle, and finally the average reverse gate power, $P_{GR(AVE)}$, is the maximum allowable reverse gate power that can be safely dissipated over a complete cycle.

Phase Control
SCR
2N4361/2N4371 Series

70 A Avg.
Up to 1400 Volts

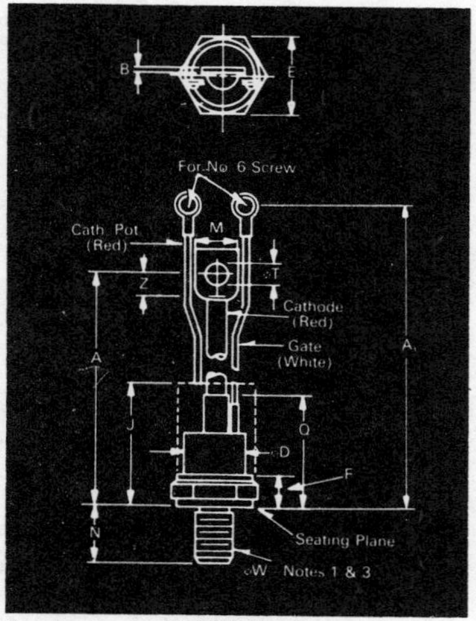

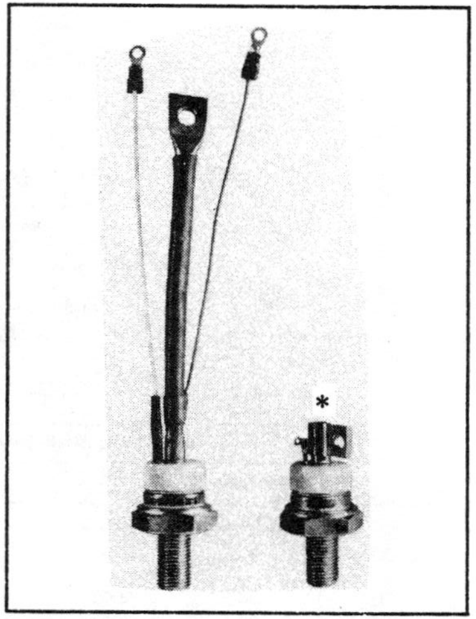

* For TO-83 Outline, see page S23.

Symbol	Inches Min.	Max.	Millimeters Min.	Max.
A	5.775	6.265	146.69	159.13
A₁	6.850	7.500	173.99	190.50
B	.055	.075	1.40	1.91
φD	.860	1.000	21.84	25.40
E	1.031	1.063	26.19	27.00
F	.255	.400	6.48	10.16
J	2.50		63.50	
M	.437	.650	11.10	16.51
N	.796	.827	20.24	21.01
Q		1.675		42.55
φT	.260	.291	6.60	7.39
Z	.250		6.35	
φW	½-20 UNF-2A			

Creep & Strike Distance.
.50 in. min. (12.85 mm).
.10 in. min. (2.54 mm). **
(In accordance with NEMA standards.)
Finish—Nickel Plate.
Approx. Weight—5 oz. (142 g).
1. Complete threads to extend to within 2½ threads of seating plane.
2. Angular orientation of terminals is undefined.
3. Pitch diameter of ½-20 UNF-2A (coated) threads (ASA B1.1-1960).
4. Dimension "J" denotes seated height with leads bent at right angles.

Applications:
- Phase control
- Power supplies
- Motor control
- Light dimmers

Conforms to TO-94 Outline

Features:
- All diffused design
- Low gate current
- Low V_TM
- Compression Bonded Encapsulation
- Low Thermal Impedance

FIG. 2-4 Westinghouse type 2N4361/2N4371 series, silicon controlled rectifier, maximum ratings and characteristics.

Voltage

Blocking State Maximums ① (TJ = 125°C)

	Symbol	2N4361 2N4371	2N4362 2N4372	2N4363 2N4373	2N4364 2N4374	2N4365 2N4375	2N4366 2N4376	2N4367 2N4377	2N4368* 2N4378
Repetitive peak forward blocking voltage, V	V_{DRM}	100	200	400	600	800	1000	1200	1400
Repetive peak reverse voltage, V	V_{RRM}	100	200	400	600	800	1000	1200	1400
Non-repetitive transient peak reverse voltage, t≤5 msec, V	V_{RSM}	200	300	500	700	950	1200	1450	1700
Forward leakage current, mA peak	I_{DRM}	10	→						
Reverse leakage current, mA peak	I_{RRM}	10	→						

Current

Conducting State Maximums (TJ = 125°C)

	Symbol	
RMS forward current, A	$I_{T(rms)}$	110
Ave. forward current, A	$I_{T(av)}$	70
One-half cycle surge current, A ①	I_{TSM}	1600
3 cycle surge current, A ①	I_{TSM}	1250
10 cycle surge current, A ①	I_{TSM}	1080
I²t for fusing (for times 8.3 ms) A²sec	I^2t	10,700
Forward voltage drop at ITM = 500A and TJ = 25°C,V	I_{TM}	2.5

Switching

(TJ = 25°C)

	Symbol	
Typical turn-off time, IT = 50A, TJ = 125°C, diT/dt = 5 A/μsec, reapplied dv/dt = 20V/μsec, linear to 0.8 VDRM, μsec	t_q	100
Typ. turn-on-time, IT = 100A, VD = 100V, μsec	t_{on}	4
Min critical dv/dt exponential to VDRM T 125 C, V μsec ①	dv/dt	100
Min di/dt non repetitive, A μsec,	di/dt	800

① ④ ⑤

Gate

Maximum Parameters (TJ 25 C)

	Symbol	
Gate current to trigger at VD 12V, mA	I_{GT}	250
Gate voltage to trigger at VD 12V, V	V_{GT}	3
Non-triggering gate voltage TJ 125 C, and rated VDRM V	V_{GDM}	0.15
Peak forward gate current, A	I_{GTM}	4
Peak reverse gate voltage, V	V_{GRM}	5
Peak gate power, Watts	P_{GM}	15
Average gate power, Watts	$P_{G(av)}$	3

Thermal and Mechanical

	Symbol		
Min., Max. oper. junction temp. °C	T_J	-40 to +125	-40 to +150
Min. Max. storage temp. °C	T_{stg}		
Max. mounting torque. ① in lb		130	
Max. Thermal resistance① Junction to case, °C/Watt	$R_{\theta JC}$	.28	
Case to sink, lubricated °C/Watt	$R_{\theta CS}$	12	

① Consult recommended mounting procedures.
② Applies for zero or negative gate bias.
③ Per JEDEC RS-397, 5.2.2.1.
④ With recommended gate drive.
⑤ Higher dv/dt ratings available, consult factory.
⑥ Per JEDEC standard RS-397, 5.2.2.6.
*2N4361 Series in TO-94 PKG.
2N4371 Series in TO-83 PKG.
**Glass-to-metal seal package.

FIG. 2-4 (cont'd.)

2-3-3   Current Parameters

The total forward current that an SCR can safely carry is dependent upon the maximum junction temperature, the thermal impedances of the device, and the losses within the device which are primarily on-state losses, provided that the device is being operated at less than the specified di/dt. Normally these ratings are based on switching frequencies up to 400 Hz and for specified conduction conditions.

The forward voltage drop V_F (V_{TM}), which is the instantaneous voltage drop between the anode and cathode under load current conditions with the SCR operating under pulsed conditions and properly heat sinked, must be equal to or less than the maximum value specified in the data sheet. A plot of V_F vs. I_F, (V_{TM} vs. I_{TM}) is shown in Fig. 2-5(e). From this plot curves of power dissipation vs. forward current for rectangular and half-wave sinusoid waveforms at various conduction angles are developed [see Figs. 2-5(a) and (c)]. In turn, this data, combined with the thermal impedances, steady-state and transient, are used to present case temperature vs. forward current for both rectangular and half-wave sinusoid waveforms [see Figs. 2-5(b) and (d)].

2-3-4   Thermal Parameters

The SCR is designed to operate with an operating junction temperature T_J within the specified range; for the 2N4361/2N4371 Series SCR this range is from $-40°C$. to $+125°C$. To maintain operation within this range it is necessary that the forward current not produce excessive junction temperatures. The maximum value of RMS current I_{RMS}, ($I_{T(RMS)}$) that the SCR may conduct is determined by the heating effects produced by:

1. The forward conduction loss, the major loss.
2. The forward or reverse leakage losses during blocking.
3. The losses produced by turn-on or turn-off.
4. The gate junction losses which are dependent upon the type and duration of the gate signal, i.e., single-pulse, pulse trains or dc signals.

The thermal resistances of a heat-sink-mounted SCR are best illustrated by considering the electrical analog of the device as shown in Fig. 2-6, where P_{AVE} is assumed to be a constant current source; then the junction temperature T_J is

$$T_J = P_{AVE}(\Theta_{jc} + \Theta_{cs} + \Theta_{sa}) + T_A \qquad (2-5)$$

THYRISTOR

70 A Avg.
Up to 1400 Volts

Phase Control
SCR
2N4361/2N4371 Series

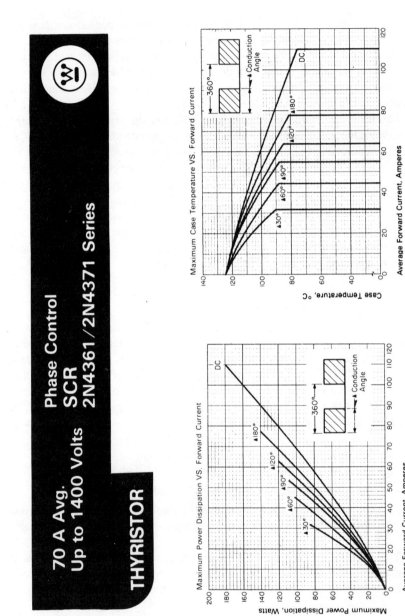

Maximum Power Dissipation VS. Forward Current

(a)

Maximum Case Temperature VS. Forward Current

(b)

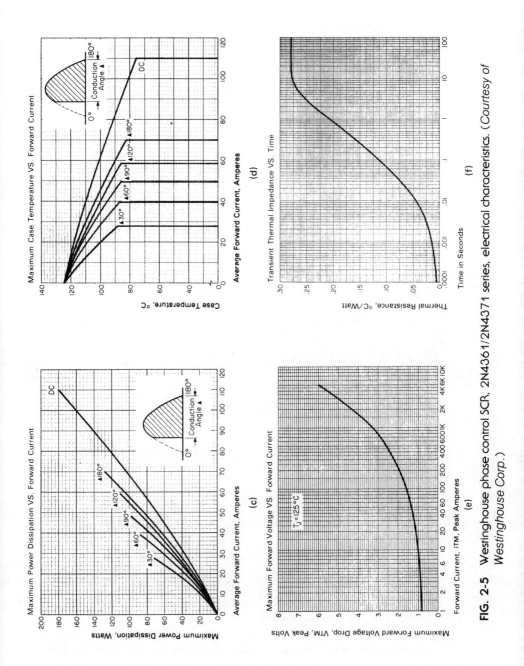

FIG. 2-5 Westinghouse phase control SCR, 2N4361/2N4371 series, electrical characteristics. (*Courtesy of Westinghouse Corp.*)

33

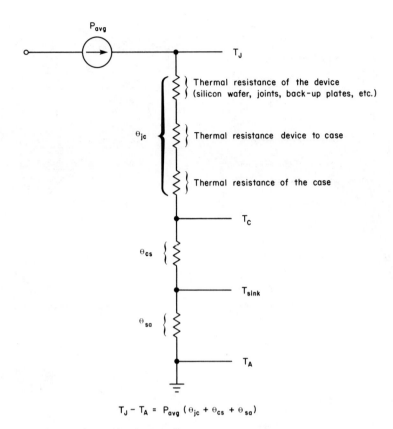

FIG. 2-6 Electrical analog describing the relationships between P_{AVE} and junction temperatures.

where

$$T_J = \text{junction temperature, °C (125°C for 2N4361/2N4371 series)}$$

$$P_{AVE} = \text{average power dissipation in the SCR, watts}$$

$$\Theta_{jc} (R_{\Theta jc}) = \text{thermal resistance, junction to case, °C/W}$$

$$\Theta_{cs} (R_{\Theta cs}) = \text{thermal resistance, case to sink, °C/W}$$

$$\Theta_{sa} = \text{thermal resistance, sink to ambient, °C/W}$$

$$T_A = \text{ambient temperature, °C}$$

The thermal resistance Θ_{sa} of the sink to ambient depends upon the type of material, shape, treatment of the outside surface, and dimensions of the heat

sink. Usually the information is presented in the form of graphs by the manufacturer, and relates the difference in temperature between the heat sink and the
ambient temperature ΔT, the average power dissipated P_{AVE} taking the form
shown in Fig. 2-7, where

$$\Delta T = T_{SINK} - T_A \ {}^\circ C \qquad (2\text{-}6)$$

and
$$\Theta_{sa} = \frac{\Delta T}{P_{AVE}} \ {}^\circ C/W \qquad (2\text{-}7)$$

To date the assumption has been that a dc current is flowing and Eqs. (2-
5) through (2-7) are applicable to steady-state conditions. Recognition must be
made of the effects of rectangular pulses of current, chopper control applications,
and half-wave sinusoids of current as would be experienced in phase angle
control.

The thermal capacity of the thyristor is very small compared to that of the
load and source, and as a result the junction temperatures of the thyristor will
vary rapidly and may exceed their designed maximum values, under cyclic load
conditions.

Consider the application of a train of rectangular power pulses as shown
in Fig. 2-8. The junction temperature T_J rises exponentially during the duration
of the pulse and then decays exponentially during the off period. The difference
of temperature ΔT is

$$\Delta T = P\Theta_{(t)} \ {}^\circ C \qquad (2\text{-}8)$$

where $\quad \Theta_{(t)} \ (Z_{\Theta(t)})$ = the thermal transient impedance ${}^\circ C/W$

A plot of the thermal transient impedance vs. pulse duration is prepared
by the manufacturer and forms a part of the SCR data sheet [see Fig. 2-5(f)].
It can be seen that the thermal transient impedance will reach a steady-state
value, the dc thermal impedance, only after a pulse of approximately 100 seconds in duration. Expressed in terms of the instantaneous forward current i_A,
with short duration pulses will result in T_J exceeding its specified value unless
the average forward current $I_{AVE} \ (I_{T(AV)})$ is less than when a constant dc current
is being carried (see Fig. 2-5(b)).

When half-wave sinusoids of current are being considered, the dc rating
of the thyristor is even greater because of the form factor of the pulse, resulting
in a higher peak current than for a rectangular pulse; curves illustrating this
situation are shown in Fig. 2-5(c).

If the current pulses are irregular, they are approximated to a regular pulse
train in order to simplify the calculations.

The foregoing is best illustrated by an example.

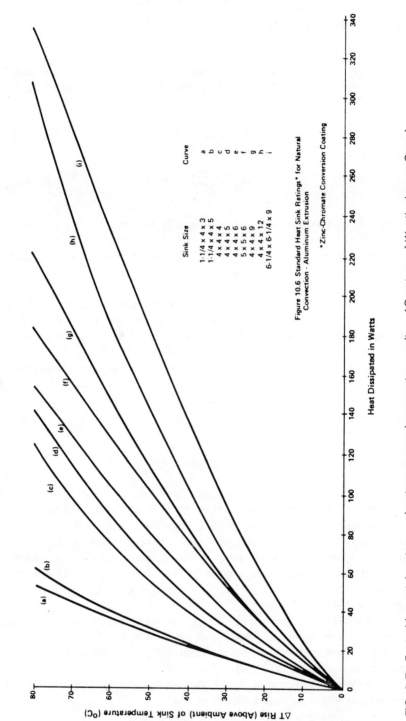

FIG. 2-7 Typical heat sink ratings—aluminum natural convection cooling. (*Courtesy of Westinghouse Corp.*)

Figure 10.6 Standard Heat Sink Ratings* for Natural Convection - Aluminum Extrusion

Sink Size	Curve
1-1/4 x 4 x 3	a
1-1/4 x 4 x 5	b
4 x 4 x 4	c
4 x 4 x 5	d
4 x 4 x 6	e
5 x 5 x 6	f
4 x 4 x 9	g
4 x 4 x 12	h
6-1/4 x 6-1/4 x 9	i

*Zinc-Chromate Conversion Coating

ΔT Rise (Above Ambient) of Sink Temperature (°C)

Heat Dissipated in Watts

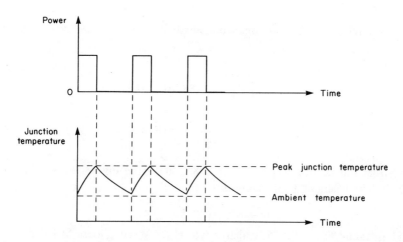

FIG. 2-8   Junction temperature variations resulting from a train of rectangular power pulses.

EXAMPLE 2-1

In Fig. 2-9, $V = 208$ V, $R = 2\Omega$, and the ambient temperature $T_A = 40°C$. Using a Westinghouse 2N4361/2N4371 Series SCR, determine: (a) the correct heat sink from Fig. 2-7 when the firing delay angle $\alpha = 0°$, and (b) the case and junction temperatures when $\alpha = 120°$.

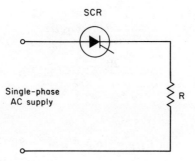

FIG. 2-9   Circuit for Example 2-1.

Solution

(a) When $\alpha = 0°$, the conduction angle $\gamma = 180°$ and the average current is

$$\frac{1}{2\pi} \int_0^\pi \frac{V_m}{R} \sin \omega t \, d(\omega t)$$

$$\frac{V_m}{R\pi} = \frac{\sqrt{2} \times 208}{2\pi} = 46.8 \text{ A}$$

and from Fig. 2-5(d), $T_{C(MAX)}$ °C $= 108°C$ and from Fig. 2-5(c) $P_{AVE(MAX)} = 76$ W. The heat sink temperature is

$$T_{SINK} = T_C - P_{AVE} \cdot \Theta_{cs}$$

and assuming $\Theta_{cs} = 0.075$, which is typical for Westinghouse SCRs, then

$$\therefore T_{SINK} = 108 - 76 \times 0.075 = 102.43 \approx 103°C$$

$$\therefore \Delta T = T_{SINK} - T_A = 103 - 40 = 63°C$$

From Fig. 2-7, curve c defines the correct heat sink for natural convection cooling that will dissipate 76 W when $\Delta T = 63°C$.

(b) When $\alpha = 120°$, the conduction angle $\gamma = 60°$ and the average current is

$$\frac{1}{2\pi} \int_\alpha^\pi \frac{V_m}{R} \sin \omega t \, d(\omega t)$$

$$= \frac{1}{2\pi} \int_\alpha^\pi \frac{\sqrt{2V}}{R} \sin \omega t \, d(\omega t)$$

$$= \frac{V}{\sqrt{2}\pi R} (1 + \cos \alpha)$$

$$= \frac{208}{\sqrt{2}\pi \times 2 \times 2} = 11.7 \text{ A}$$

From Fig. 2-5(c), $P_{AVE} = 16.5$ W

From Fig. 2-7, curve c for $P_{AVE} = 16.5$ W, $\Delta T = 20°C$; therefore,

$$T_{SINK} = T_A + \Delta T = 40 + 20 = 60°C$$

and the case temperature T_C is

$$T_C = T_{SINK} + P_{AVE}\Theta_{cs} = 60 + 16.5 \times 0.075$$

$$\simeq 61°C$$

and the temperature of the junction T_J is

$$T_J = T_C + P_{AVE}\Theta_{jc} = 61 + 16.5 \times 0.28$$

$$= 64.6°C$$

$$\simeq 65°C$$

2-3-5   Additional Current Parameters

Recognition must be taken of current conditions that affect temperature such as partial use of the cathode area and high current density operation. These conditions result from fuse clearing, circuit breaker opening, fault currents and high switching rates. These current conditions result in junction overheating.

2-3-5-1   RMS Current, I_{RMS}, $(I_{T(RMS)})$

The RMS current is used to rate the device, but in most applications the average current delivered to the load is the one of most importance, and as a result the manufacturer prepares his data in terms of the average current I_{AVE}, $(I_{T(AV)})$. Because the form factor (the ratio of RMS/AVE), varies with the conduction angle for half-wave sinusoids, resulting in I_{RMS} increasing as the conduction angle decreases, assuming a constant I_{AVE}, I_{RMS} is responsible for producing the heat losses in the device and increasing temperatures.

2-3-5-2 I^2t Ratings

The I^2t rating is used to define the thermal capacity of fuses, and in the protection of thyristors the I^2t rating of the fuse is selected to be less than the I^2t rating of the thyristor. These ratings are usually based on the fuse clearing a fault in less than half a cycle, but prior to clearing, the thyristor will be subject to excessive temperatures because the device is acting as a resistance. The specified I^2t rating for the thyristor is that value necessary to enable selection of the correct protection without damaging overheating occurring at the junction.

2-3-5-3   Surge Current $I_{FM}, (I_{TSM})$

The surge current, $I_{FM}, (I_{TSM})$, which is the peak value of a single half-cycle current pulse at 60 Hz, is nonrepetitive, and the device is designed to withstand a maximum of 100 surges during its lifetime, since the maximum junction temperature is exceeded with a resulting loss of blocking capability and increase in the leakage currents. Permissible values are specified as 1600 A for one-half cycle, 1250 A for three cycles, and 1080 A for 10 cycles, for the 2N4361/ 2N4371 Series SCR.

   In normal operation the junction temperature may be at its maximum rated value; then if a surge of current occurs under this condition, the maximum T_J will be exceeded. If there is insufficient time after the application of a surge, the device may not be able to block the off-state voltage.

2-4   DYNAMIC CHARACTERISTICS

The static anode-cathode characteristics do not present any information regarding the turn-on and turn-off times. This information is of vital interest when forced commutation techniques are used in chopper control and static frequency conversion.

2-4-1   Turn-on Time

The application of a gate pulse to the gate of an SCR does not result in the immediate flow of anode current. Initially, there is no significant increase in the anode current, this interval is known as the *delay time* t_d, which is defined as the time interval between a specified point at the beginning of the gate pulse and the instant when the device voltage (current) has dropped (risen) to 90 percent of $V_{FB}, (V_{DRM})$ (10 percent of the final value of I_A). Similarly, the rise time t_r is the time interval between when the anode-cathode voltage has decreased from 90 percent to 10 percent of its original value, or the anode current has increased from 10 percent to 90 percent of its final value. The turn-on time, t_{on}, is equal to $(t_d + t_r)$ and is usually defined in the data sheets. The relationship among I_A, I_g, and V_{AK} is shown in Fig. 2-10.

   Normally the rise time t_r is sufficiently small at normal power frequencies, i.e., 60 Hz and 400 Hz, that the SCR is in full conduction before the peak of the applied anode-cathode voltage is reached. In the case of high di/dt applications, initially conduction will be limited to a small area of the silicon wafer in the immediate vicinity of the gate connection, and localized heating may occur with the possibility of device failure.

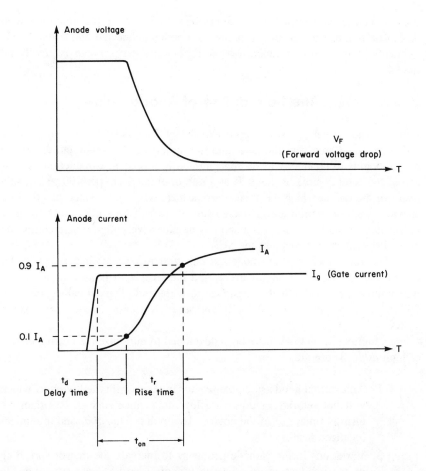

FIG. 2-10 SCR turn-on definitions in terms of anode voltage and current for a pure resistive load.

There is also a relationship between t_d and t_r of the SCR and the rise time and amplitude of the initiating gate pulse. In high *di/dt* applications the device delay and rise times can be decreased by using a gate pulse with a fast rise time, typically between 0.1 and 1 μsec, with a peak amplitude 3.5 to 5 times the minimum gate current required to achieve turn-on, and with a pulse duration of at least 10 μsec. In the case of the 2N4361/2N4371 Series SCR t_{on} is 4 μsec. When the SCR is used in a "hard" circuit, a pure resistive circuit, the duration of the rise time is very important, since there will be a significant forward voltage and current present simultaneously; and the instantaneous power and thus the heat developed in the device can be significant. In this situation the di/dt can be reduced to acceptable limits by a series connected anode reactor. In applications involving inductive or "soft" circuits the problem is greatly reduced.

2-4-2 di/dt, The Rate of Rise of Anode Current

Initially when a gate pulse is applied to the SCR, conduction is limited to a relatively small area near the gate, and the initial flow of anode current as the device begins to conduct will be concentrated in this area. Conduction spreads across the cathode area of the SCR at a rate of about 1 cm per 100 μsec. As a result, if the rate at which the anode current increases is far greater than the rate at which the conduction area is increasing, there will be a high power density in this area, resulting in local hot spots with excessively high temperatures and possibly permanent damage to the SCR.

The rate of change of anode current with respect to time is called the critical rate of rise of on-state current, and the term di/dt is used to define the maximum di/dt that the SCR is designed to withstand. Typical values are 30 to 200 A/μsec for phase control SCRs and as high as 800 A/μsec for inverter SCRs.

A number of methods have been developed to minimize the heating effects of high di/dt. These are:

1. An external anode saturable reactor in series with the thyristor, which will not saturate or go into a low impedance state in less than the turn-on time, t_{on}, of the device. This reduces the di/dt and minimizes localized heating.

2. Variations in the cathode geometry to increase the rate of spreading of the conduction area by the use of improved gate structures. In general, the types of gate structures used by Westinghouse are:

    a. *Edge fired:* Originally the gate was placed at the periphery of the cathode for ease of manufacture, which is still the basic design for most low di/dt light industrial SCRs [see Fig. 2-11(a)].

    b. *Center fired:* The disadvantage of the edge fired design was a relatively low rate of spreading of the conduction area. By moving the gate connection to the center of the structure the rate of spreading was greatly improved [see Fig. 2-11(b)].

    c. *Power integrated circuit:* With an increase in the required switching rates, an improved design was required to permit the SCR to be used in high frequency applications. The power integrated circuit SCR was developed to meet this requirement. It consists

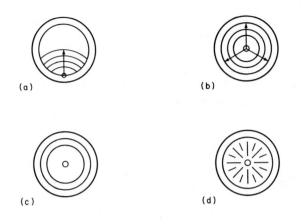

FIG. 2-11 Cathode geometries used to increase the di/dt capabilities of SCRs. (a) edge fired; (b) center fired; (c) power integrated circuit; (d) interdigitated.

basically of a pilot thyristor which is used to turn on the main SCR, the whole combination being formed on the one silicon wafer [see Fig. 2-11(c)].

d. *Interdigitated cathode:* The periphery of the gate was greatly increased, resulting in a more rapid rate of increase of the conduction area [see Fig. 2-11(d)].

2-4-3 Critical Rate of Rise of Forward Voltage, *dv/dt*

In inverter and cycloconverter applications, the thyristor is subjected to an extremely fast rising wave front which causes a charging current $i = Cdv/dt$ to be produced, which when added to the normal leakage current will initiate turn-on. This current is a function of the junction capacities, the blocking voltage, and the rate of application of the anode voltage, *dv/dt*. Other sources of the critical rate of rise of forward voltage are ac switching surges and transients on the ac supply; also, operation at elevated temperatures will reduce the *dv/dt* capability of the thyristor, since less gate current is required to achieve turn-on.

The shorted emitter type of construction is one commonly used method to increase the *dv/dt* capability of a thyristor. This is accomplished by providing a shunt resistance path around the gate-cathode junction, so that when the voltage across this resistance exceeds the gate-to-cathode voltage a junction current flows and turn-on is achieved. The *dv/dt* ratings in data sheets are given for the device operating at maximum junction temperature and with a zero gate signal. Typical values of *dv/dt* are 25 to 300 V/μsec for phase-controlled SCRs and as high as 800 V/μsec for inverter SCRs.

By itself, *dv/dt* turn-on is not destructive to the device, but in some applications such as an inverter it could create a short circuit across the dc source. There are several ways of minimizing the problem, one of which is to use an R-C snubber network in parallel with the SCR, where the capacitor absorbs the excess transient energy and the resistor limits the current (see Fig. 2-12)

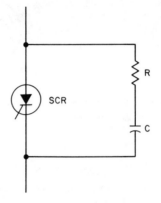

FIG. 2-12 *dv/dt* suppression, R-C snubber network.

Another effective approach is to select a thyristor with a higher *dv/dt* rating, or a thyristor with a higher voltage rating, since $V_{FB}, (V_{DRM})$ will be greater; then the ratio of the applied anode voltage to $V_{FB}, (V_{DRM})$ will be lower and the *dv/dt* capability of the thyristor will be greater.

Alternatively, applying a reverse-bias to the gate will result in an increase in the forward breakover voltage and the *dv/dt* capability; however, care must be taken to ensure that the negative bias does not exceed V_{GRM} as specified in the data sheets. A side benefit of negative gate biasing is that it reduces the possibility of transient turn-on caused by noise signals in the gate circuit.

2-4-4   Dynamic Turn-off

In the conducting state each of the three junctions are forward-biased, and the base regions are saturated as depicted in Fig. 2-13. Before forward blocking can be achieved the charge carriers must be removed. One method, which is not practical, is to open the anode circuit. Normally, however, the anode-to-cathode voltage is reduced until the holding current can no longer be maintained, at which point the charge carriers recombine and the thyristor regains its forward

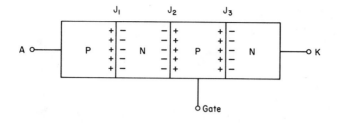

FIG. 2-13 SCR in conduction.

blocking capability. This process is called commutation and occurs naturally in ac applications. In dc applications since the anode-to-cathode does not become reverse-biased naturally, it is necessary to apply a reverse-bias to reduce the anode current by external means. This process is called forced commutation.

When the anode-to-cathode voltage becomes reverse-biased, the anode current decreases to zero and becomes negative, reaching a peak value which is called the reverse recovery current $I_{R(REC)}$. This current causes electrons and holes to migrate away from the J_2 junction to the J_1 and J_3 junctions, until the carriers have recombined at the J_2 junction. Up to the point where the reverse recovery current peaks, there is still a small positive voltage drop across the thyristor. At this point the anode-to-cathode voltage becomes reverse-biased and increases to its maximum value; at the same time the anode current decreases to zero. The time interval between the point at which the reverse recovery current commences to the point where it becomes zero is called the reverse recovery time t_{rr}. This time interval is of the order of 2 μsec for SCRs of less than 100 A, and can increase to the order of 6 to 8 μsec for high-current SCRs.

In high-voltage rating SCRs the silicon wafer is usually thicker to prevent a voltage breakdown of the device, and as a result the recombination time at the J_2 junction increases, with a resultant increase in t_{rr}.

The forward blocking junction requires a period of time called the gate recovery time, t_{gr}, to establish a depletion region before the forward blocking voltage can be re-established. The gate recovery time is usually much longer than the reverse recovery time. The period of time $(t_{rr} + t_{gr}) = t_{off}$, is known as the circuit commutated turn-off time, and is usually specified in the SCR data sheets; it can vary from 20 μsec for inverter SCRs to the order of 150 to 200 μsec for phase-controlled SCRs. The waveforms showing the voltage and currents during turn-off are shown in Fig. 2-14.

The duration of the turn-off time, t_{off}, is affected by the following:

1. The amplitude of the anode current prior to commutation; for anode currents less than the designed current, the turn-off time will be faster,

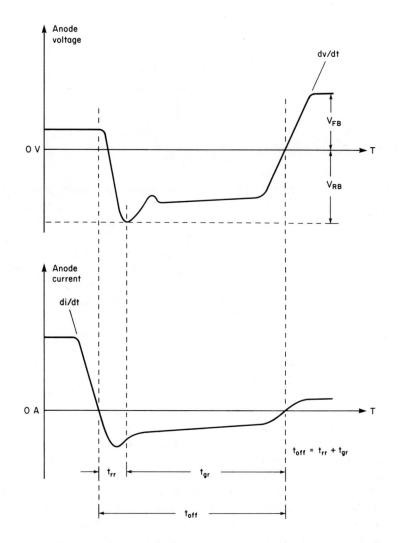

FIG. 2-14 Voltage and current waveforms during turn-off.

and conversely for rated currents, if it is assumed that the device is
being operated at rated junction temperatures.

2. The junction temperature, T_J; at elevated anode currents that result
   in junction overheating, the turn-off time will be increased.

3. The $-di/dt$ of the on-state current; the rate of change of current is governed by the external inductance of the circuit—the more rapid the rate of change of current, the shorter will be the reverse recovery time and the total turn-off time will be reduced.

4. The amplitude of the reverse voltage; reverse voltage promotes the reverse recovery current, and thus variations in the amplitude of the reverse voltage directly affect the rate of change of the reverse recovery current. Reverse voltages of the order of 25 to 30 volts during the gate recovery time, t_{gr}, decrease the turn-off time.

5. The rate of rise dv/dt of the reapplied forward voltage; the rate of reapplied forward voltage dv/dt must be less than that specified in the t_{off} specification. For the 2N4361/2N4371 Series SCR it is 20 V/μsec and is significantly less than the critical dv/dt of 100 V/μsec.

It should also be appreciated that the turn-off time available to the thyristor should be greater than the specified turn-off time, in order that the forward blocking capability may be achieved. This is not normally a problem at power line frequencies, but in high frequency applications the turn-off time can be a significant proportion of the total time-available and recourse must be made to fast switching SCRs.

2-5   GATE CHARACTERISTICS

Most SCR manufacturers provide recommended values of V_{GT} and I_{GT}, either in the form of a graph (Fig. 2-15) or in the form of tables, that may be applied to the SCR to achieve turn-on without exceeding the maximum gate power P_{GM} that can be safely dissipated.

In Fig. 2-15 curves 1 and 4 are the limiting diode characteristics, line 2 defines the maximum permissible gate voltage, and line 3 represents the maximum permissible gate power P_{GM}. The trigger circuit to ensure reliable turn-on must provide gate signals (V_{GT} and I_{GT}) that fall within the area bounded by lines 1, 2, 3, and 4. The small area at the origin of the graph represents all values of $V_{GNT}, (V_{GD})$ and $I_{GNT}, (I_{GD})$ that are guaranteed not to achieve turn-on.

The determination of suitable values of V_{GT} and I_{GT} can be made by load line analysis. The load line is constructed by drawing a straight line between the trigger source open circuit voltage on the ordinate, and the short circuit current on the abscissa. The load line is drawn for two different values of source impedance R_G, where $R_G = V_{GT}/I_{GT}$. In the first case, $R_G = 5$ ohms which is too small since it crosses the maximum gate power dissipation curve; in the second case, $R_G = 20$ ohms. It can be seen that in general, low impedance sources are required for gate trigger circuits.

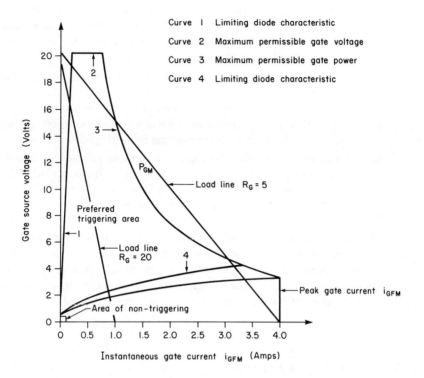

Curve 1   Limiting diode characteristic

Curve 2   Maximum permissible gate voltage

Curve 3   Maximum permissible gate power

Curve 4   Limiting diode characteristic

FIG. 2-15 Typical gate characteristics.

It can also be seen from Fig. 2-15 that the maximum values of V_{GT} and I_{GT} can be far in excess of those required to guarantee turn-on.

In the case of low load di/dt conditions, the selection of the gate drive circuit is solely decided by V_{GT} and I_{GT} values selected by load line analysis. However, in high di/dt applications, such as those that would be experienced in inverter and chopper applications, a hard drive gate signal is necessary to initiate conduction and promote rapid spreading of the conduction area. It is usual practice to apply a pulse that will peak at $5.0\ I_{GT(MAX)}$ at the end of the gate rise time t_r; after approximately 5 μsec the gate current is reduced to $I_{GT(MAX)}$. Typical values for the 2N4361/2N4371 Series SCR are $i_{gt(MA)} = 333$ mA, and $V_{GT} = 4.3$ V at $T_J = 25°$C.

In order to minimize gate-cathode junction overheating it is normal not to apply a continuous dc pulse for the period that the anode-cathode is forward-biased, but rather a pulse no longer than 100 μsec or alternatively a 5- or 10-

kHz pulse train such that the average gate power $P_{G(AVE)}$, $(P_{G(AV)})$ over the complete cycle is not exceeded.

2-6 GATE FIRING CIRCUITS

Three basic types of gate firing signals are usually used. These are dc signals, pulse signals and ac phase signals. However it is not intended to give an exhaustive treatment at this point; more specific examples will be dealt with in Chapter 8.

2-6-1 DC Signals

As previously mentioned, it is not desirable to apply a continuous dc gate signal because of the excessive gate power dissipation that will be experienced. Normally dc gating signals are not used for triggering SCRs in ac applications, since during the negative half-cycle the presence of a positive gate signal would increase the reverse-anode current with possible destruction of the device. Figure 2-16(a) illustrates a simple method of applying a dc signal from an external

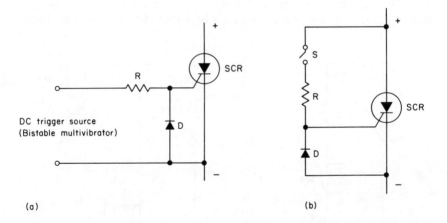

(a)   (b)

FIG. 2-16 DC gating signals. (a) from an independent source; (b) from a derived source.

trigger circuit, which can take many forms ranging from a simple single-pole switch to a bistable multivibrator. The function of the resistor R is to limit the gate current, and the diode D limits the amplitude of a negative gate signal to approximately 1 volt. Figure 2-16(b) shows an alternative method of providing the gate signal without requiring an external source. In both cases the circuits operate in basically the same manner.

2-6-2 Pulse Signals

The major advantages of using pulse signals, either single or multiple, are that the power dissipation at the gate-cathode junction is reduced, and electrical isolation between the gating signals and the device is possible. The advantage of using electrical isolation, by means of pulse transformers or optocouplers, is that a number of thyristors may be gated from the same source, and it permits the minimization of spurious noise signals, such as transient noise spikes causing inadvertent operation of the thyristor.

The most common method of producing pulses is by the use of the unijunction transistor (UJT) relaxation oscillator, as illustrated in Fig. 2-17(a). This circuit provides a train of narrow pulses with a slow rise time. The pulse repetition rate of the pulses is determined by the R-C time constant of $(R1 + R2)C$, and when the emitter voltage V_E of the UJT reaches the peak value, the emitter-base 1 junction goes into a low impedance state, and emitter current flows through the primary of the gate pulse transformer, applying the gate signal to the SCR. An inherent problem with this circuit, because of the narrow pulse width, is that a latching current may not be achieved before the gate signal is

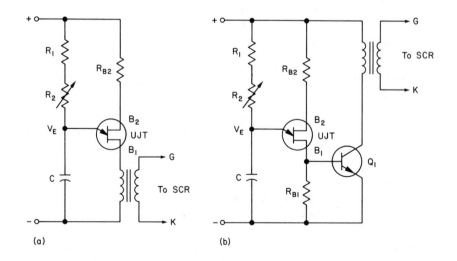

FIG. 2-17 UJT relaxation oscillator; (a) basic circuit; (b) with improved pulse width and rise time.

removed. This disadvantage may be eliminated by using an R-C snubber network across the SCR.

The width and rise time of the pulse may be improved by using the output across R_{B1} to drive a transistor amplifier, as shown in Fig. 2-17(b). The pulse width of the output signal can be increased by increasing the value of C.

2-6-3 AC Signals

By far the most common method of controlling thyristors in ac applications is to derive the firing signal from the ac source, and to control its point of application to the thyristor so that the firing delay angle may be varied during the positive half-cycle. The techniques can vary from a simple resistive control, as illustrated in Fig. 2-18(a), where variation of the potentiometer $R2$ will delay

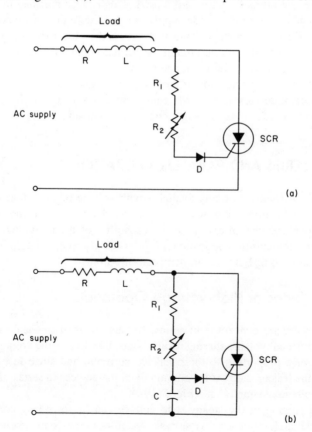

FIG. 2-18 AC signals; (a) resistive control; (b) R-C phase control.

the firing of the thyristor by controlling the rise of the gate voltage. This arrangement is feasible when changes due to temperature and di/dt are low; however, the maximum variation of the firing delay angle is limited to 0° to 90°. An improved ac control is obtained by using an R-C delay circuit to produce the gating signal, but once again the range of firing delay angle control is limited to 0° to 90°. One precaution that must be taken is to ensure that the R-C network charging current is in excess of I_{GT}, and that when $R2$ is at its maximum value, I_{GT} still will be produced.

2-7   SYNCHRONIZATION

In ac applications, single- or three-phase, it is essential that the gating signal be applied when the thyristor is forward-biased. This is accomplished by synchronizing the gate circuit to the anode supply, so that the timing cycle commences at the point when the anode voltage is crossing the zero axis going positive, and, in the case of three-phase circuits, generates trains of gate pulses spaced 120° apart with respect to the reference point.

To achieve synchronization of the trigger currents to the ac source, usually the trigger circuit derives its power from the same ac source, and by phase shift control techniques, firing delay angle control is obtained.

2-8   SERIES AND PARALLEL OPERATION

Even though SCRs are currently available with voltage ratings of several thousand volts, applications exist where it is necessary to operate the devices in series to block voltages in excess of the capability of the individual thyristor. Similarly, in high-current applications, it is necessary to operate the devices in parallel in order to operate within their current ratings.

2-8-1   Series or High-voltage Operation

When thyristors are connected in series, because of their unequal static characteristics, unequal voltage sharing results, since they divide the voltage between them in inverse proportion to their leakage currents, and since in a series arrangement the leakage current is common to all the devices in series, the devices share the voltage unequally (see Fig. 2-19).

Equal sharing of the steady-state voltage can be forced by connecting a resistor in parallel with each thyristor, the value of the resistance being selected so that a current in excess of the leakage current flows through the resistance. See Fig. 2-20(a).

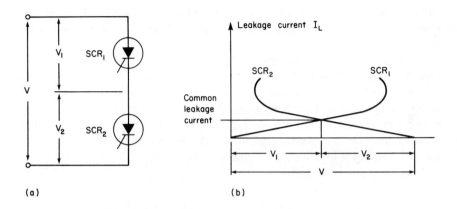

(a)   (b)

FIG. 2-19 Steady-state voltage sharing between thyristors in series. (a) basic circuit; (b) static characteristics.

During turn-on the slowest thyristor to turn on will have the full source voltage applied across it, and similarly during turn-off the fastest thyristor to turn off will be subjected to the full source voltage. Under transient turn-on and turn-off conditions it is necessary to promote equal transient voltage sharing. The transient voltage distribution across the thyristors is in inverse proportion to their capacitances. Transient voltage sharing is achieved by shunting each thyristor by a capacitor greater in value than the thyristor capacitance [see Fig. 2-20(a)]. In order to reduce the surge of current from this capacitor through the thyristor during turn-on a resistance is connected in series with each capacitor.

The remaining problem to be solved is the simultaneous turn-on of all seriesed devices. First, there must be electrical isolation between the thyristor

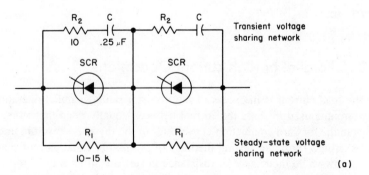

FIG. 2-20 Voltage sharing in seriesed thyristors. (a) transient and steady-state sharing; (b) simultaneous gating.

gates, which are considerably above ground or neutral potential, and the gating circuits, which are low voltage circuits. Second, the last device to be turned on is subjected to excessive voltage stress, which can lead to the destruction of the device.

The trigger circuit must provide a fast rising pulse (less than 1 μsec) at least 50 μsec wide with current overshoot, to ensure turn-on even with an inductive load. The usual arrangement is to use either a pulse transformer with a common primary and multiple secondary windings or to use optocouplers to provide the electrical isolation [see Fig. 2-20(b)]. The resistors $R3$ are used to equalize the gate impedances and to reduce the possibility of current being shunted away by a low impedance gate circuit.

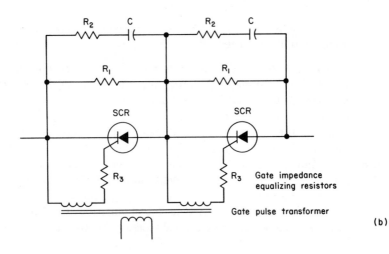

FIG. 2-20   (cont.)

2-8-2   Parallel or High-current Operation

When the load current is in excess of the thyristor rating, parallel operation of the thyristors is used to share the current between equally rated thyristors. Differences in the forward conduction capabilities of the thyristors result in unequal current sharing. There are several methods of promoting equal current sharing, such as a low value inductance or resistance in series with each thyristor, or by a current sharing reactor, as illustrated in Fig. 2-21. This is basically a center-tapped reactor in which the thyristor carrying the greater current will cause an unbalance in the ampere-turns across its half of the reactor, which in turn will induce an aiding voltage that will cause an increase of current flow through the thyristor which originally carried the lower current, thus achieving a balance in current distribution.

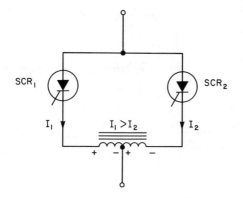

FIG. 2-21 Paralleled thyristors, forced current sharing.

During turn-on, the fastest thyristor will be carrying all of the current, unless a hard gate signal is applied and maintained until all the paralleled thyristors have latched.

The basic rules that determine the required number of paralleled thyristors are:

1. The magnitude of the continuous load current.

2. The magnitude of the current during a short-term overload duty cycle.

3. The magnitude and duration of a fault current.

Since the prime problem in high-current applications is excessive junction temperatures, the current carrying capacity of the thyristors is derated, and the thyristors are usually mounted on a common heat sink to equalize temperatures.

To achieve satisfactory latching, a fast rise overdriven gate pulse of sufficient duration must be applied simultaneously to all the paralleled thyristors (see Fig. 2-22). The function of the resistance in series with each gate is to reduce the effects of shunting by a low impedance gate.

When high-voltage, high-current applications are met, series-parallel combinations of thyristors are used.

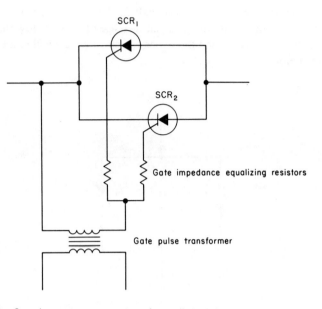

FIG. 2-22 Simultaneous triggering of paralleled thyristors.

2-9   PROTECTION

To obtain satisfactory operation of thyristors and their associated equipment, it
is necessary to provide protection against high voltage transients and overcurrent. The protection problem can be very complex and may be best illustrated
by considering an ac-supplied, phase-controlled converter for four-quadrant control of a dc motor. This system combines heavy rotating equipment, power
handling thyristors and diodes, and low-power control components and circuits.

2-9-1   Voltage Transients

Voltage transients are primarily caused by switching disturbances, the source
usually being the stored energy ($\frac{1}{2}LI^2$) in the inductive components of the system. The effect is to produce voltage transients whose peak may be as much as
ten times the repetitive forward blocking voltage V_{FB}. In the forward direction
anode turn-on may occur, possibly causing the flow of large fault currents; in
the reverse direction the transient can cause large currents to flow in small areas
around the thyristor junction and can produce localized heating, unless transient
suppression techniques are applied.

Voltage transients in the sample system can be broadly classified as ac side transients, dc side transients, and transients occurring within the rectifying equipment.

2-9-1-1   Transients Originating on the AC Side of the Rectifier

These transients are caused by:

1. Line voltage switching resulting from fault clearance and switching in the ac supply system. The magnitude of the transient is dependent upon the speed of switching and the point on the voltage sinusoid at which switching occurs.

2. Lightning. It produces a line-to-ground surge with a steep-fronted waveform, building up to a maximum in a few microseconds, and it is completed in a few hundred microseconds.

3. Transformer primary inrush current. It produces oscillations in the resonant secondary winding circuit caused by the transformer leakage reactance and the distributed secondary winding capacitance. Applying the primary voltage at its peak produces a secondary voltage transient approximately twice the normal peak value. Even higher transients are produced if contact bounce is present.

4. Interruption of the transformer magnetization current. The rapid decay of core flux induces a transient at the secondary terminals which may approach ten times the peak secondary voltage. The transient is greatest at no load, when the connected rectifier is supplying an inductive load, for example, a choke input filter. The amplitude will be at its greatest if the interruption occurs when the primary current is passing through its zero value.

5. Energizing the primary of a step-down transformer. Because of the interwinding capacitance between the primary and secondary windings, when the primary is energized the primary voltage is momentarily coupled to the secondary. In high-ratio, step-down rectifier transformers the interwinding capacitance is of the order of 0.001 microfarads, and the charging voltage can give rise to transient secondary voltages several times the normal secondary voltage.

With reasonable care, the elimination of the sources of these voltage transients can be achieved. For example, switching the secondary of a transformer will overcome the problem of the oscillations caused by the primary inrush current. Transients caused by fault clearance in the ac supply system can be minimized by ensuring that the interrupting devices have a long arc time, thus dissipating the transient in the arc. Lighting arresters placed near the equipment to be protected will bypass overvoltages to ground. Connecting a capacitor across the secondary of a step-down transformer will cause voltage division and a reduction of the transient.

When it is not practical to suppress transients by these means, then R-C snubbers across the input ac lines may be used, or alternatively, voltage dependent symmetrical resistors such as General Electric Thyrectors and GE-MOV varistors may be used.

2-9-1-2   Transients Originating on the DC Side of the Rectifier

When dc side switching takes place under load conditions, the stored magnetic energy in the ac supply system and the transformer leakage reactances produce a voltage -Ldi/dt. If there is no alternative leakage path, the induced voltage then appears across the SCRs and the switch, and it will be at its greatest under dc short-circuit conditions.

Inductive loads on the dc side will not cause voltage transients when switched, since by Lenz's Law the induced voltage will maintain the foreward-biased SCRs in conduction.

Under overhauling load conditions the armature of a dc motor is accelerated, and the counter-emf is increased above the rectifier output level and applies a high reverse voltage across the SCRs. Similar conditions exist if the shunt field excitation is rapidly increased while the motor is driving a high inertia load.

The reduction of dc side generated transients is achieved by dissipation of the stored energy across the rectifier output terminals. Some methods are:

1. Adding additional inductance in series with the dc load.
2. Using a controlled rate of opening switch (or fuse for dc side fault clearance).
3. Connecting thyrectors or varistors across the rectifier output.
4. With regenerative loads, connecting a resistance in series with the armature when the voltage exceeds a safe value.

2-9-1-3   Transients Occurring Within the Rectifying Equipment

The principal causes of transients generated within the rectifier are the interruption of semiconductor fuses under fault conditions, and commutation transients.

1. Interruption of semiconductor fuses. Under fault conditions the rate of change of fault current is di/dt and is produced by the applied circuit voltage V, where $V = Ldi/dt$, L being the circuit inductance, at the end of the melting time $di/dt = 0$. The voltage across the arc is V. For di/dt to decay at the same rate as the rise, the voltage across

the arc must be two times V. This voltage is applied across the semiconductor device, and unless the voltage rating of the device is sufficiently high, it will cause device failure.

2. **Commutation transients.** After turn-off has been initiated, time is required for the charge carriers to be swept away from the junctions, before the thyristor is capable of blocking a reverse voltage. During this period the reverse recovery current is limited by the circuit impedance. When blocking is achieved, the sudden termination of the reverse recovery current causes a voltage transient to be generated by the commutating inductance.

2-9-2   Overcurrent Protection

In overcurrent protection, the nature of the source impedance is very important. If the source impedance is "soft"—that is, there is sufficient reactance to limit the di/dt of the circuit—and its duration is unlikely to be in excess of a few cycles of the source frequency, then sufficient protection of the thyristor can be obtained by selecting a thyristor with an adequate surge current rating. If, on the other hand, the duration of the fault current is more than a few cycles, then it may be sufficient to rely upon a conventional short-time delay electromagnetic relay or thermal overloads, provided that prior to the clearance of the fault the thyristor does not become overheated.

If the source impedance is "stiff"—that is, it is a low-reactance source—then a rapid current buildup will occur under fault conditions. In this case, it is essential that the circuit be interrupted rapidly before permanent damage can occur to the thyristor. The cheapest and most effective interrupting device in these circumstances is the semiconductor fuse, which will clear the circuit when:

1. Another thyristor fails to block in the reverse direction, or
2. A short circuit occurs across the output terminals of the converter, or
3. A thyristor fails to turn off in time, or turns on at the incorrect time.

The action of a fuse under fault conditions is illustrated in Fig. 2-23. Initially, under fault conditions the current rises rapidly, only being limited by the circuit impedance. If there were no fuse action, it would build up to a peak value as shown and decay to zero, if it is assumed that a half-wave sinusoid voltage is being applied. In actual practice the current builds up to point A, at which point the fuse begins to melt. There will be a further slight increase in the current magnitude up to point B, the peak let-through current, before the energy is dissipated in the arc, with complete interruption of the protected circuit being achieved at point C.

The total clearing time, which is designed to occur in less than 8.3 msec,

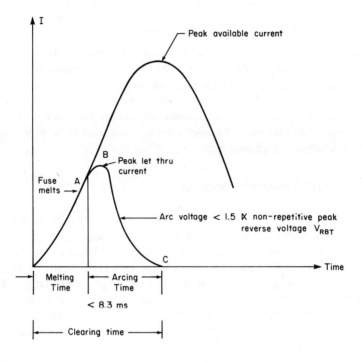

FIG. 2-23 The protective action of a semiconductor fuse.

consists of two equal time segments, the melting time and the arcing time. The rate of decrease of current during the arcing time must also be low enough that high induced voltages (*Ldi/dt*), which could destroy the thyristors, are not produced.

Semiconductors and fuses are designed to withstand specified I^2t values below a period of one-half cycle of the supply voltage.

2-9-2-1 I^2t Ratings

The energy that must be dissipated by a fuse during the clearing time comes from two sources, that from the power supply, and that from the stored magnetic energy ($\frac{1}{2}LI^2$) in the inductive portions of the circuit. The heat energy to be dissipated by the fuse is $\int i^2R \, dt$, where R represents the circuit resistance. Since the current is common to the whole circuit and the resistance is basically constant during the clearing time, the energy to be dissipated is proportional to $\int i^2 \, dt$, i.e., I^2t (amperes^2 × time).

By rating the fuses and the semiconductors on a common basis, namely their I^2t ratings, it is then possible to select fuses with a lower I^2t rating than the semiconductor, and thus to protect the semiconductor.

Fuse and thyristor manufacturers specify in their data the I^2t ratings of their devices, usually based on total clearing times of the fuses not exceeding 8.3 msec. The I^2t rating of the thyristor is based on the device operating at maximum rated current and maximum junction temperature; since this is not the normal practice, a factor of safety is introduced.

Since semiconductor fuses are much more expensive than normal renewable link fuses, it is normal practice to consider the relative costs of the fuse against the cost of the semiconductor and its replacement, and in the case of low cost thyristors to eliminate fusing.

2-9-3   Gate Circuit Protection

The gate circuits must also be provided with protection because of inductive and capacitive coupling between power and control circuits inducing voltage transients into the gating circuitry. Because of the high di/dt in the power circuits, it is essential that the power wiring and control wiring be as widely separated as possible, which also has the side effect of reducing the capacitive coupling.

Electrical isolation is standard practice in high-power converters and is usually provided by gate pulse transformers, which are ferrite cored, with a primary and usually multiple secondaries, or by optocouplers. Usually the leads connecting the gate and cathode of the SCR to either the pulse transformer or optocoupler are twisted with a minimum of two twists per foot, or are shielded, with the shield being connected to a common at one end only.

Figure 2-24 illustrates a number of gate terminations which are designed to discriminate between signals and noise. In Fig. 2-24(a) the resistor decreases the gate sensitivity, increases the dv/dt of the device, and reduces the turn-off time, which at the same time increases the holding and latching currents. Connecting a capacitor between the gate and cathode leads [Fig. 2-24(b)] removes high-frequency noise components and increases the dv/dt capability of the thyristor; at the same time, it increases the gate-controlled delay time t_d and turn-on time and turn-off time, and reduces the gate-controlled rise time t_r. The circuit of Fig. 2-24(c) protects the gate supply during reverse transients in excess of V_{RBT}, and limits the negative bias applied to the gate to approximately 1 volt; alternatively, a zener diode may be used. Figure 2-24(d) gives protection in low-power thyristor applications with dc gate signals. The capacitor C removes the ac component of a transient noise signal, and the resistor improves the dv/dt capability of the device. These gate terminations are only representative of the many combinations that can be used.

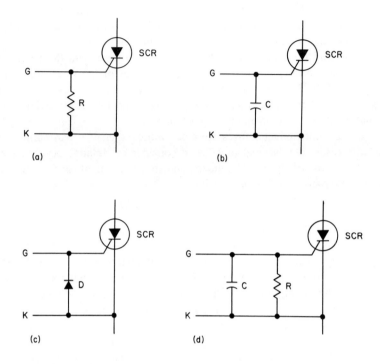

FIG. 2-24 Gate terminations.

2-10   THE BIDIRECTIONAL TRIODE THYRISTOR—TRIAC

The TRIAC overcomes one of the major objections to the use of the SCR in ac phase control; it can conduct in both directions and can be controlled by a positive or negative gate signal.

The TRIAC is basically equivalent to two SCRs connected in inverse-parallel on the same silicon wafer. The basic structure is shown in Fig. 2-25(c). The inverse-parallel SCRs are gated on by applying a gate signal to SCR1, when MT1 is negative, and a gate signal to SCR2 when MT2 is negative, an arrangement which requires two separate gating sources [see Fig. 2-25(a)].

The TRIAC differs from this arrangement in that the two gates are connected together and a single gate connection is brought out [Figs. 2-25(b) and (c)].

The static switching characteristic is shown in Fig. 2-26. The characteristic in quadrant I is identical to that of the SCR, and since effectively the TRIAC is an inverse-parallel arrangement of SCRs, the characteristic in quadrant III is symmetrical to that in quadrant I.

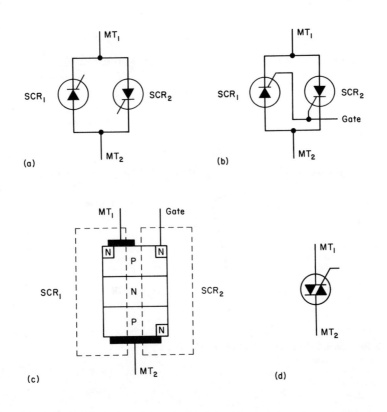

FIG. 2-25 (a) Two SCRs in anti-parallel; (b) with a common gate connection; (c) TRIAC pellet structure; (d) circuit symbol.

Normally the TRIAC may be turned on by low-power positive or negative gate signals in quadrants I and III, with the gate signal being applied between the gate and the MT1 terminal.

The four operating modes are as shown in Table 2-2.

TABLE 2-2

Gate to MT1 Voltage	MT2 to MT1 Voltage	Operating Quadrant
Positive	Positive	$I(+)(+I_{GT}, +V_{GT})$
Negative	Positive	$I(-)(-I_{GT}, -V_{GT})$
Positive	Negative	$III(+)(+I_{GT}, +V_{GT})$
Negative	Negative	$III(-)(-I_{GT}, -V_{GT})$

Typical values of gate currents for an RCA 40668, 8 A, 120 V TRIAC are $I(+)$ 10 mA, $III(-)$ 15 mA, $I(-)$ 20 mA, and $III(+)$ 32 mA at a case temperature of 20°C.

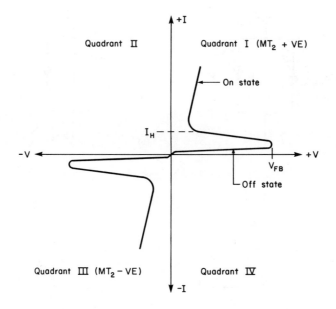

FIG. 2-26 Static V-I characteristics of a TRIAC.

From these gate current figures, it can be seen that the TRIAC is not equally sensitive to the gate signals in all four modes, being least sensitive to III(+) signals. It can be deliberately made to be inoperative in the III(+) mode, and it will operate with a negative gate in either direction, but with a positive gate signal it will act as an SCR. When operating in this manner the device is called a logic TRIAC.

Since the TRIAC can conduct in both halves of the ac cycle, current ratings are specified in the data sheets for a conduction period of 360 deg.

In the TRIAC, because of its capability of conducting in both halves of the ac cycle, it may have been conducting immediately prior to blocking in the other direction, and it must be able to support the commutating dv/dt. In the case of inductive loads, the TRIAC turns off when the load current is zero; however, the voltage across the TRIAC rises very rapidly to the instantaneous ac supply voltage, and as a result the TRIAC is subjected to a high dv/dt, which may be in excess of the capability of the device. Under these conditions the device may be subjected to dv/dt turn-on, which can be minimized by an R-C snubber network across the device. If this is not successful, then the inverse-parallel arrangement of SCRs must be used to control highly inductive loads.

2-11   SUMMARY

A good deal of attention has been given to the operating characteristics of the SCR in order that its limitations and capabilities are understood when it is applied in subsequent chapters. In addition, it is quite often necessary to make substitutions in the course of maintenance; by understanding the SCR data sheets, satisfactory substitutions can be made.

REVIEW QUESTIONS

1. Explain the operation of the thyristor in terms of the two-transistor analogy.
2. Explain the means by which thyristor turn-on can occur.
3. Explain the significance of the rms current in rating an SCR.
4. What is the significance of the I^2t ratings of fuses and thyristors?
5. What is the significance of the term di/dt, and how can its effects be reduced?
6. What is meant by the critical rate of rise of forward voltage, dv/dt, and how can its destructive effects be minimized?
7. What parameters affect the turn-off time t_{off} of a thyristor?
8. Discuss high-voltage and high-current application of SCRs and explain the protection circuits that must be provided to prevent device failure.
9. Analyze the causes and effects of voltage transients originating on the ac and dc sides of a rectifier and those originating within the rectifier. How may they be minimized?
10. In what way does a semiconductor fuse differ from a renewable type? Under what conditions must the fuse clear the circuit?

3 Static Control

3-1 INTRODUCTION

The bistable properties of thyristors, that is, conducting and nonconducting, are leading to an ever-increasing application in the field of static switching of control and power circuits. However, the refined design of mechanical and electromechanical circuit-interrupting devices, combined with their low cost, has presented the thyristor with a stiff challenge in meeting the competition.

The thyristor has a number of advantages as compared to other switching devices:

1. No contact bounce upon closing; this minimizes transients.
2. Silent operation, since there are no moving parts.
3. With the use of zero voltage turn-on techniques and the zero current turn-off characteristic of the thyristor, it practically eliminates radio frequency interference (RFI).
4. Negligible maintenance, since there are no contacts or moving parts to replace. The life expectancy is of the order of 20 million operations versus approximately 1 million by an electromechanical switch.
5. Completely safe in explosive atmospheres.
6. Vibration and shock immune.
7. Can be mounted in any position or location.

8. Very high switching speeds.

9. Small and lightweight.

10. Readily adaptable to electronic control.

11. Lower cost.

Some of the disadvantages are:

1. Liable to fail under overvoltage and overcurrent conditions unless suitably protected.

2. In the nonconducting state it does not provide complete electrical isolation from the supply source because of leakage currents.

3. In some applications the 0.5–1.5 volt drop across the device may not be acceptable.

4. Usually restricted to single-pole applications, mainly because of the cost of the devices and the complexity of the control.

5. An SCR static switch can be designed for ac or dc operation, but not for both applications in the same switch, since the ac switch is line-commutated and the dc switch is force-commutated.

6. They are subject to false turn-on by high dv/dt caused by voltage transients from other switching devices in the supply. This problem may be reduced by high-frequency filtering of the supply voltage.

7. Triggering circuits must be shielded, or be twisted pairs isolated from power circuits to prevent false turn-on.

8. Surge current and fault current detection circuits must be capable of turning off the device before the surge current and I^2t ratings are exceeded.

Static switching using thyristors, in addition to the normal on-off action of conventional circuit-interrupting devices, can be designed to provide time-delay, latching, over and undercurrent and voltage detection, stepping and selector switch action. The gating signals can originate from transducers detecting mechanical, electrical, light, position, proximity, etc., signals.

Static switching applications fall into two categories: ac switching and dc switching. In the case of ac switching, since the device is line-commutated, the upper frequency limit is determined by the type of device. In the case of the TRIAC, the normal range is from 50 to 400 Hz, and for the SCR up to approximately 30 kHz. In dc switching, the switching speed is determined by the limitations of the forced commutation circuitry and the recovery time of the SCR.

3-2   STATIC AC SWITCHES

There are basically three sections to the ac switch: namely, the power, the triggering, and the protection circuits.

3-2-1   Inverse-parallel SCR Connection and TRIAC Connection

This is the simplest full-wave, single-phase ac contactor configuration. From 400 Hz to 30 kHz the inverse-parallel SCR configuration is used. Up to 400 Hz, provided the current rating is not exceeded, a TRIAC may be used [see Fig. 3-1(a) and (b)]. To overcome the effects of transients in the ac supply, an $R\text{-}C$ snubber network, especially with an inductive load, should be connected in parallel with the thyristors.

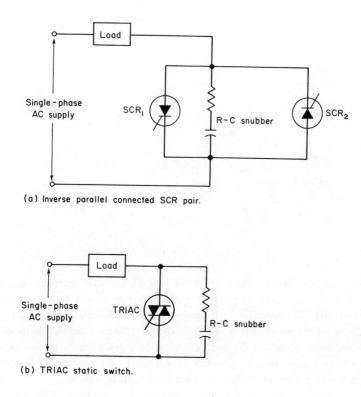

(a) Inverse parallel connected SCR pair.

(b) TRIAC static switch.

FIG. 3-1   Single-phase ac static contactor.

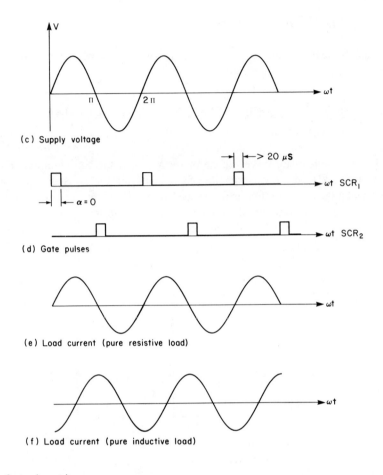

(c) Supply voltage

(d) Gate pulses

(e) Load current (pure resistive load)

(f) Load current (pure inductive load)

FIG. 3-1 (*cont.*)

To minimize distortion of the load waveforms, the thyristors must be turned on at the beginning of the half-cycle of the ac supply voltage [see Fig. 3-1 (c) and (d)]. The resultant load current waveforms for a pure resistive and pure inductive load are shown in Fig. 3-1 (e) and (f). In the case of an inductive load, a gate pulse signal of 20 μsec duration may not be sufficient to achieve turn-on. In this case an *R-C* network in parallel with the load will produce sufficient latching current for the thyristor to remain conducting.

The inverse-parallel SCR cathodes are usually grounded to the case, and as a result the case of one of the SCRs must be insulated from ground by an insulator that is a good thermal conductor.

The forward blocking voltage V_{FB} must be at a minimum equal to the peak line voltage, that is, $\sqrt{2} V_L$. For 220-V circuits the peak voltage is 311 V, and safe practice would be to use a 500-V thyristor to provide protection against line transients.

If the instantaneous line current is $I_m \sin \omega t$, then the RMS line current is

$$I_{RMS} = \frac{1}{\pi} \int_0^\pi I_m^2 \sin^2 \omega t \, d(\omega t) \tag{3-1}$$

and since each thyristor carries the line current for only one half-cycle, the average current carried by each thyristor is

$$I_{AVE} = \frac{1}{2\pi} \int_0^\pi I_m \sin \omega t \, d(\omega t) = I_m/\pi$$

Substituting for I_m from Eq. (3-1), then we have

$$I_{AVE} = (\sqrt{2} I_{rms})/\pi = 0.45 I_{rms} \tag{3-2}$$

As a result it is necessary to select SCRs so that their current rating is at least 0.45 times the load current.

3-2-2 Alternative Static Switching of Full-wave Single-phase Power

Figure 3-2 illustrates some variations using combinations of thyristors and diodes to obtain full-wave static ac switching of single-phase circuits. In Fig. 3-2(a), control of both halves of the ac cycle is obtained by connecting the thyristor across the dc terminals of the full-wave dc bridge. The current through the load is ac, and that through the thyristor is dc. The load may be resistive or inductive. However, if turn-off of the thyristor after the removal of the gate signal is to be achieved when the dc voltage drops to zero, the inductance of the thyristor circuit must be very small. In the case of an inductive load it should be shunted with an R-C snubber circuit to compensate for the phase shift caused by the inductive load.

In Fig. 3-2(b), SCR1 and diode D2 conduct for one half-cycle, and SCR2 and D1 conduct for the other half-cycle, but in this case the gate firing control has a common ground, since the thyristor cathodes are common.

Another switching alternative utilizing auxiliary thyristors is shown in Fig. 3-2(c). The auxiliary thyristors SCR1 and SCR4 are low-current types and maintain a constant gate signal on the main thyristors SCR2 and SCR3. In order to maintain the gate dissipation of the main thyristors within designed limits, a limiting resistance R is required in series with the auxiliary thyristors. Typical values of R would be in the range of 1.1 k for 110 V operation and 2.2 k for 208 V operation, based on the assumption of 500 mW gate dissipation with a gate current of 100 mA. Providing a lower voltage in phase with the supply voltage to the auxiliary thyristors will also reduce the gate dissipation.

Figure 3-2(d) shows one pole of an ac bus transfer switch, whose function is to provide switching between a preferred and alternate ac source to an important load such as a computer. With this arrangement it is possible to achieve changeover within one-quarter cycle of an interruption of the preferred source, which is impossible with an electromechanical device. The arrangement is for

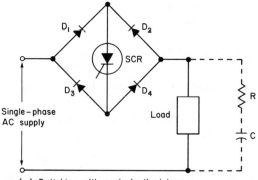

(a) Switching with a single thyristor

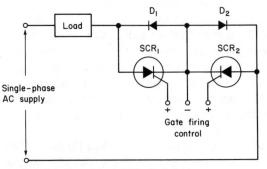

(b) Switching with an SCR-diode bridge

FIG. 3-2 Single-phase ac static switches.

one pole only in each source. Each pole consists of a pair of inverse-parallel-connected SCRs; under normal operation the load is supplied from the preferred source. The voltage sensor detects any voltage decrease below the value deter-

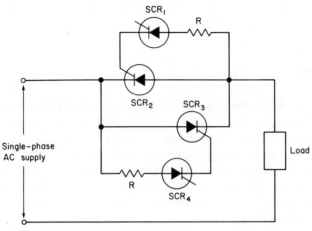

(c) Switching with auxiliary thyristors

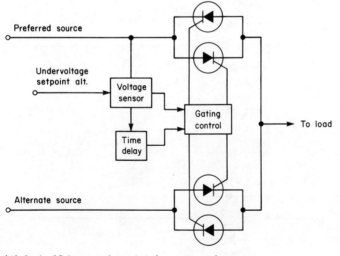

(d) Static AC bus transfer switch (one pole only)

FIG. 3-2 *(cont.)*

mined by the under-voltage setpoint adjustment, at which point the conducting SCRs are turned off and the alternate source inverse-parallel-connected SCRs are gated on. On restoration of the preferred supply for in excess of the interval determined by the time delay, control is transferred back to the preferred supply thyristors. The function of the time delay is to ensure that a stable preferred supply is available before switching over from the alternate source.

3-2-3   Three-phase Static Contactors

The concept of single-phase ac static switching is readily extended to three-phase applications, as shown in Fig. 3-3. Basically inverse-parallel-connected thyristors are inserted in series with the ac supply lines of a star or delta-connected load. In order to reduce costs it is permissible to substitute a diode for a thyristor in each of the inverse-parallel-connected circuits of Fig. 3-3(a) and (b). Further cost reductions may be made, provided that unbalanced voltage operation is acceptable, by eliminating one set of inverse-parallel-connected thyristors as in Fig. 3-3(c).

A reversal of the three-phase power supplied to the load can be achieved by the circuit shown in Fig. 3-4. In Fig. 3-4, with SCRs 7 to 10 inclusive off, line A feeds $T1$, line B feeds $T2$, and line C feeds $T3$. Turning off SCRs 3

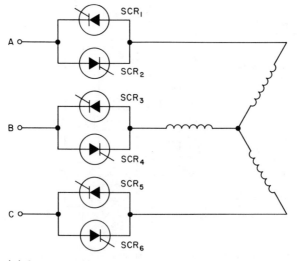

(a) Star connected load — 3-phase control

FIG. 3-3 Three-phase ac static contactors.

to 6 inclusive and turning SCRs 7 to 10 inclusive on results in line *A* feeding *T*1, line *B* feeding *T*3, and line *C* feeding *T*2, thus achieving a phase reversal of the voltages supplied to the load. It should be noted that in this configuration all active devices must be thyristors, since if diodes were used, phase-to-phase shorts would occur.

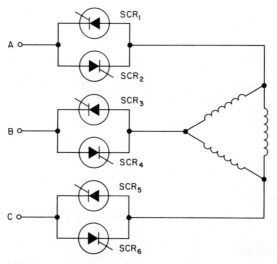

(b) Delta connected load — 3-phase control

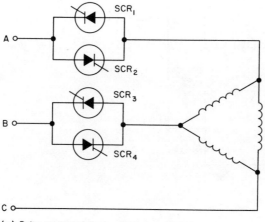

(c) Delta connected load — 2-phase control

FIG. 3-3   (*cont.*)

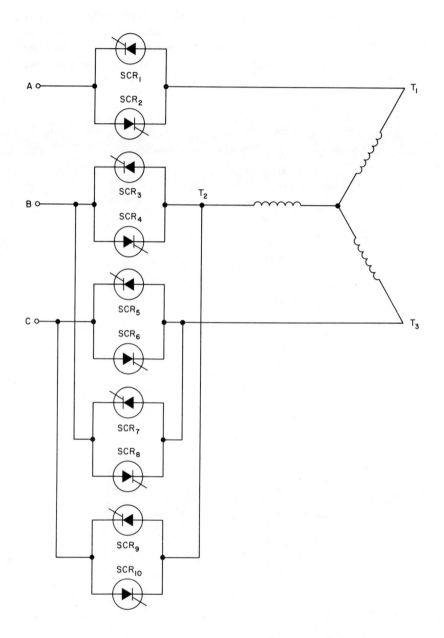

FIG. 3-4 Inverse-parallel connected three-phase reversing contactor.

3-2-4   Triggering Circuits

The triggering of the thyristors in Fig. 3-1 for a pure resistive or slightly inductive load is satisfied by applying trigger pulses of approximately 20 μsec duration at the point when the anode-cathode voltage of the respective thyristor is going positive. In the case of a highly inductive load, the phase angle ϕ between the voltage and current becomes greater than the width of the trigger pulse, and the thyristors will not be gated. This problem can be overcome by applying a continuous gating signal for the period $(180° - \alpha)/\omega$ sec, where α is the firing delay angle. But because of the need to isolate the gating signals of the thyristors, it is usual to provide electrical isolation by means of a gate pulse transformer (usually with 1:1 ratio) or a photocoupler. The physical size of the transformer increases with the pulse duration, and as a result it is more common to use a train of short pulses for the required intervals, namely, $\alpha < \gamma < 180°$ for SCR1 and $\alpha + 180° < \gamma < 360°$ for SCR2. The frequency of these pulses usually is of the order of 10 kHz (see Fig. 3-5), which prevents saturation of the pulse transformer.

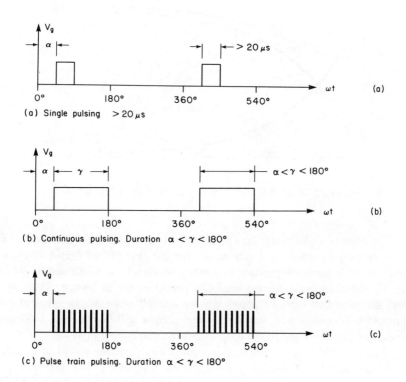

(a) Single pulsing $> 20\,\mu s$

(b) Continuous pulsing. Duration $\alpha < \gamma < 180°$

(c) Pulse train pulsing. Duration $\alpha < \gamma < 180°$

FIG. 3-5 Types of gate firing signals.

The problem of ensuring that the thyristor is gated correctly is compounded by the presence of an Ldi/dt voltage, which affects the thyristor in two ways. First, the voltage may exceed the device dv/dt and cause false turn-on, and second, it reduces the time available for the device to turn off. If at the end of a single gate pulse, a latching current has not been achieved, then the thyristor will not conduct. This situation may be overcome by placing an R-C snubber combination in parallel with the thyristor; the instantaneous sum of the capacitor current I_C and the inductive current I_L is the current I_{SCR}, which can be made to exceed the latching current, as shown in Fig. 3-6. Suitable values of R and C are of the order of $R = 100 \ \Omega$ and C between 0.3 and 2.2 μF. The selection of R and C must also ensure that at the moment of turn-on the capacitor current is controlled so that the peak current and the di/dt of the thyristor are not exceeded.

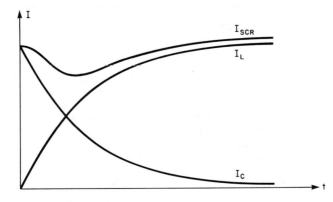

FIG. 3-6 The effect of an R-C combination paralleled across an SCR.

To operate under worst load conditions, i.e., a pure inductive load, it is necessary only to maintain a gate signal for the first 90° of each half-cycle. Therefore, it is common practice to apply gate signals to SCR1 and SCR2 of Fig. 3-1 simultaneously for the total ON period of the ac switch. The major disadvantage is that the gate pulses will be applied when one or other of the thyristors is reverse-biased, but the leakage current will be minimal, since the reverse voltage is limited to the drop across the conducting thyristor.

Figure 3-7 shows a method of providing the triggering pulses to inverse-parallel-connected SCRs utilizing a unijunction transistor (UJT) relaxation oscillator, whose frequency may be varied by $R2$ in series with C. When the UJT switches on, the capacitor discharges through the primary of the pulse trans-

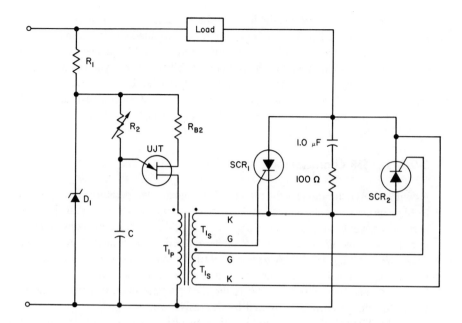

FIG. 3-7 UJT relaxation oscillator firing control for inverse-parallel connected SCRs.

former $T1$, and the dual secondary windings produce anti-phase trigger pulses for the SCRs.

3-2-5   The Solid-state Relay (SSR)

Solid-state relays (SSRs) are rapidly coming into prominence for the control of ac and dc power. In ac applications the most commonly used device is the TRIAC, and in dc applications the transistor is used.

The SSR is now available in compact encapsulated units in two basic types, all solid-state or hybrid, which require a driver to boost the signal to a reed relay. Most SSRs are electrically isolated between the control circuit and the ac load circuit; this is commonly achieved by optocouplers. In addition, the dc control types are usually compatible with digital logic devices; the ac control types usually operate over a wide input range of voltages (90 to 280 V ac), which may also be operated with dc control signals in the range of 80 to 140 V dc. Another feature that is common to ac units is the provision of zero voltage switching.

Currently, SSRs are available from the Crydom Division of International Rectifier in 120 and 240 V ac ratings with maximum load current ratings of 2.5, 10, 25 and 40 A and in a 480 V ac rating with 8 and 12 A capacity and have photo isolation, integrated circuit (IC) compatibility for low-voltage dc control signals, or ac input signal models controlling TRIACs. These devices are commonly being used for the control of motor loads, transformers, resistance heating, and high inrush lighting loads in the range of control previously covered by NEMA Size 0 and Size 1 contactors.

3-2-5-1   SSR Characteristics

As with all thyristor devices the effects of voltage and temperature are important. In high inrush current applications care must be taken to ensure that the surge current is not exceeded on a repetitive basis. In addition, since the control device is a TRIAC, the di/dt rating must not be exceeded, or localized heating will occur; di/dt ratings of 100 A/μsec are normal, and if necessary, a series-connected inductor will ensure that this rating is not exceeded. In addition to the load di/dt, another source of di/dt is the R-C snubber circuit in which the capacitor discharge current is limited by the resistance.

The TRIAC also is sensitive to turn-on resulting from voltage transients caused by switching inductive loads on adjacent lines. The effect of transients can be minimized by R-C snubbers or varactors.

The commutating dv/dt describes the capability of the TRIAC to turn off when the load current becomes zero. In inductive circuits care must be taken to ensure that the surge current rating of the SSR is not reached on a repetitive basis; however, this is not normally a serious problem, since one-cycle surge current ratings are usually about ten times normal rated current. Since in ac applications, the device is a TRIAC, the di/dt rating must not be exceeded, or localized heating will occur; however, di/dt ratings of 100 A/μsec are normal, and if necessary the addition of a series inductor will ensure that this rating is not exceeded. Another source of di/dt is from the parallel-connected R-C snubber used to provide transient voltage protection. This effect can be minimized by using a 0.1 μF capacitor in series with a 100 ohm resistor.

Normally the TRIAC has a dv/dt rating of the order of 200 to 300 V/μsec, which will give protection against dv/dt turn-on in nearly all applications. However, in the case of inductive loads, the commutating dv/dt rating must be considered. This rating describes the ability of the TRIAC to turn off when the current becomes zero. It must be appreciated that in inductive circuits the current may lag the voltage by quite an appreciable angle. As long as current is flowing, the voltage across the device is of the order of 1 volt, but when current ceases flowing the voltage across the device goes suddenly to the line voltage; this change may be of the order of 50 volts. The time required for the TRIAC to achieve its forward blocking capability is finite, and forward blocking cannot be achieved unless the commutating dv/dt is held to the order of 100 to 200 V/μsec. This can be accomplished by the use of an $R\text{-}C$ snubber network in parallel with the device.

In high ambient temperature applications the SSR requires a large heat sink or the derating of the device; additional heat sinking can quite often be obtained by mounting the SSR on the metal chassis of the cabinet.

Normally input-output isolation is obtained in SSRs by either optocouplers or transformers; however, in hybrid SSRs, a reed relay provides the isolation (see Fig. 3-8).

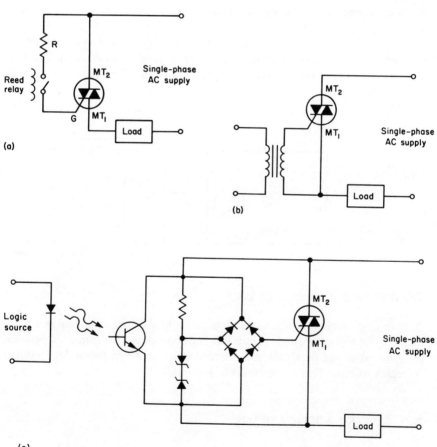

FIG. 3-8 Electrical isolation methods: (a) reed relay, (b) gate pulse transformer and (c) optocoupler.

3-2-5-2   Zero Voltage Switching (ZVS)

Experience has shown that thyristors will develop the least amount of electro-magnetic interference (EMI), if in ac applications the thyristor is turned on at the earliest possible instant after the applied voltage crosses the zero axis. Industrial standards specify that the thyristor must turn on before the anode-cathode voltage exceeds 5 volts in order to meet NEMA WD-2 EMI specifications.

Zero voltage switches, such as RCA CA3058, CA3059, and CA3079, are available as monolithic integrated circuits and are used to trigger thyristors in ac power control and switching applications. These devices detect when the ac voltage crosses the zero axis and produce an output pulse of approximately 100 μsecs duration centered on the zero crossing point; this pulse in turn is used to trigger the thyristor.

Normally SSRs are packaged for printed circuit board mounting or for heat sink mounting. The package for ac applications consists of an isolating unit, usually an optocoupler, a zero voltage switch, and a TRIAC; it is illustrated in Fig. 3-9.

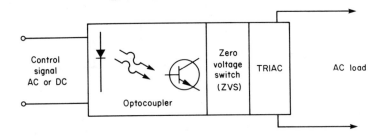

FIG. 3-9 Block diagram of an SSR.

In dc-controlled SSRs, the response time to a dc control signal is a maximum of one half-cycle, and the turn-off time corresponds to the interval between the removal of the dc signal and the next point at which the ac load current decreases to zero. This is illustrated in Fig. 3-10.

3-3   SSR APPLICATIONS

Most SSRs are supplied as the equivalent to the electromechanical single-pole-single-throw (SPST) switch, because of the costly duplication of circuitry required for multipole applications, so as a result multipole arrangements are built around this configuration. The following examples of SSR applications are by

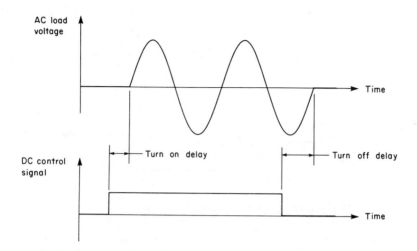

FIG. 3-10 SSR control of an ac load voltage by a dc control signal.

no means an exhaustive review but are intended to give an insight into methods of applying the SSR in areas where previously electromagnetic relays were the commonly used switching element.

Figure 3-11 illustrates the ON-OFF control of an ac load by the use of a dc control signal. In many applications the dc control is derived from 5-V TTL digital logic, and it is then necessary to sink the input of the SSR to ground via the input logic. Most manufacturers provide dc input SSRs with a range of dc input voltages, typically from 3–32 V, if the input signal is greater; then a series resistor is inserted. If the control signal is in the range of 80 to 140 V dc, then usually an ac input SSR is used instead of the dc input SSR.

Figure 3-12 illustrates the application of a dc input ac SSR and an ac input ac SSR, which are both single-pole-single-throw, to produce a single-pole-double-throw (SPDT) configuration controlling two ac loads by means of a dc input control signal.

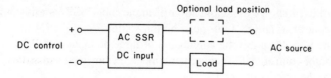

FIG. 3-11 On-off control of an ac load by means of a dc controlled SSR.

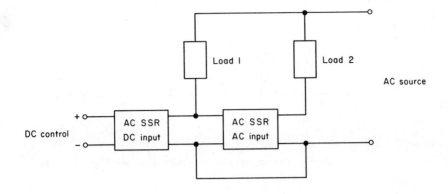

FIG. 3-12   Simultaneous control of two ac loads.

Figure 3-13 shows an application of the SSR for the control of high-power SCRs mounted in an inverse-parallel configuration for ac phase control.

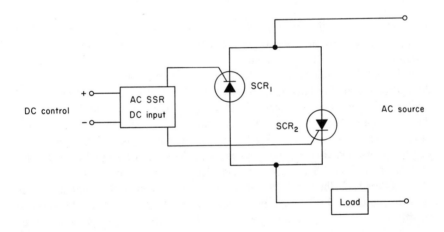

FIG. 3-13   Low-power control of high-power thyristors.

Figure 3-14 shows the application of an ac input SSR as a latching relay under the control of an ON-OFF push button station.

Figure 3-15 illustrates the application of three dc input ac SSRs to a three-phase ac motor control. *R-C* snubber networks are connected across the TRIACs to minimize the effects of *dv/dt*.

The SSR can be applied in a multitude of configurations to achieve tem-

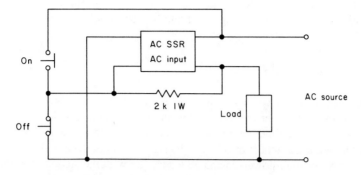

FIG. 3-14 Ac input SSR used as a latching relay.

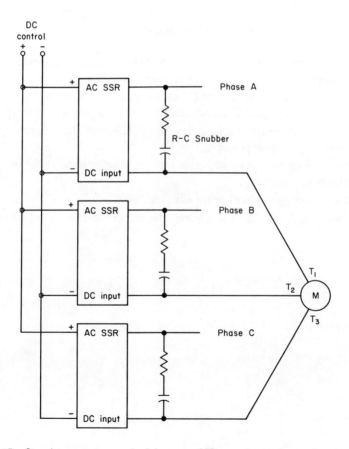

FIG. 3-15 Simultaneous control of three ac SSRs to obtain three-phase control.

perature control and sequential and adaptive systems controlling such industrial applications as batch processing, continuous processes, machine control, etc.

3-4   SUMMARY

Even though ac and dc static switching techniques and devices have a significant advantage over the older and more conventional electromagnetic control methods, it will be some time before major inroads are made by devices such as the SSR. One major advantage of SSRs is their ability to interface with digital logic in the control of high-power dc and ac loads, and with the advent of the microprocessor and minicomputer in system control, the market potential and applications of SSRs are practically infinite.

REVIEW QUESTIONS

1. What are the advantages and disadvantages of using a thyristor switch as compared to a mechanical or electromechanical switch?
2. What are the advantages and disadvantages of the inverse-parallel SCR connection and TRIAC connection methods of ac static switching?
3. Discuss with the aid of schematics the techniques of static switching of full-wave, single-phase power.
4. Discuss with the aid of schematics the application of ac static switching to three-phase contactors.
5. Discuss the principle of operation of an SSR.
6. Discuss with the aid of schematics the use of the SSR to the control of ac loads.

4 AC Voltage Control

4-1   INTRODUCTION

The power supplied to a load from an ac source can be controlled by varying
the rms value of the ac voltage by means of thyristors. There are two major
methods of control. The first is ON-OFF control, which connects the ac source
to the load for a number of cycles and then disconnects it for a similar period;
i.e., it varies the duty cycle to the load. The second is phase control, in which
the amount of power supplied over each cycle is varied by delaying the point
of conduction.

The major applications of ac voltage control techniques are industrial
heating, illumination levels, motor speed control (polyphase induction motors
or universal motors), on-load tap changing and ac magnet control.

4-2   ON-OFF CONTROL, PULSE BURST MODULATION

Pulse burst modulation control is very readily applied to industrial heating and
motor speed control because of the high thermal and mechanical inertia present.
A side benefit of pulse burst modulation combined with zero voltage switching
is to greatly reduce the radio frequency interference caused by the thyristor
switching.

In pulse burst modulation control, a fixed time period, the periodic time T, is defined in terms of a number of cycles of the ac source frequency. This time period is divided into two segments, t_{ON} and t_{OFF}. The duration of t_{ON} is usually determined by a feedback loop, which permits a comparison between the reference input and a fixed percentage of the signal representing the controlled variable, i.e., temperature or speed. The t_{ON} period usually consists of an integral number of cycles, as illustrated in Fig. 4-1.

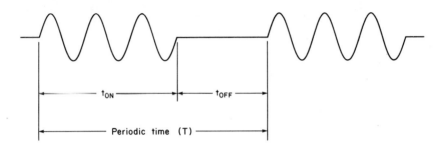

FIG. 4-1   Pulse burst modulation control.

The most common thyristor arrangements to implement this form of control are inverse-parallel-connected SCRs or a TRIAC, t_{ON} being controlled from 0 to 100 percent of the periodic time T.

4-3   AC PHASE CONTROL

There are a considerable number of circuit configurations that can be used to control the transfer of power in ac applications by means of thyristors. In applications up to 400 Hz and within the current ratings currently available, the TRIAC is the most commonly used device. In high-frequency or higher-current applications inverse-parallel-connected SCRs are used. Either configuration permits delayed firing in either half of the ac cycle.

4-3-1   Single-phase Phase Control Circuits

Some of the more common thyristor circuits used in single-phase ac phase control are discussed in the following paragraphs.

Figure 4-2 illustrates the half-wave controller and load voltage and current waveforms. This circuit is suitable for controlling low-power resistive loads, such as heating and lighting; the maximum range of control is from 50 to 100

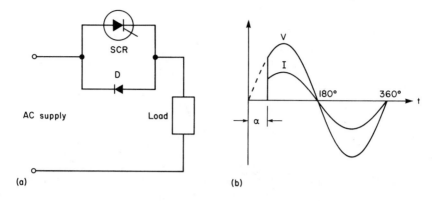

FIG. 4-2 Half-wave controller (a) basic circuit; (b) load voltage and current
waveforms resistive load.

percent because of the uncontrolled diode. The asymmetrical waveform may
introduce a dc component, which, if large enough, could cause saturation of the
core of the supply transformer.

Figure 4-3 shows the full-wave controller, which can be used to control
a resistive or a resistive-inductive load. Either SCR1 or SCR2 is in conduction
at any one time, the range of the firing delay angle control being from 0 to 180
deg with the gate pulses being synchronized to the ac line and 180 deg apart.

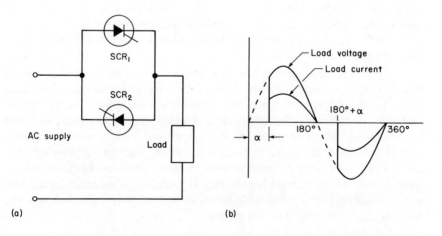

FIG. 4-3 Full-wave controller, (a) basic circuit (b) load voltage and current
waveforms, resistive load, (c) load voltage and current waveforms,
resistive-inductive load.

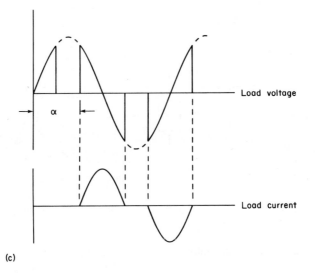

(c)

FIG. 4-3 (*cont.*)

With inductive loading it is usual to lengthen the pulse in order to achieve a latching current. With a pure resistive load, the load current will cease as soon as the SCRs become reverse-biased. However, in the case of an inductive load, which stores energy in the magnetic field, the collapsing magnetic field will maintain the SCR in conduction even after it is reverse-biased until the current decays below the holding current, at which point the load voltage will remain zero until the next SCR is turned on.

Figure 4-4 illustrates a full-wave controller, which can be used to control an ac load or a dc load. This is the least efficient form of controller, since there are always three devices in conduction at the same time. In addition, the voltage applied to the thyristor is unidirectional, and with a resistive or slightly inductive load when the voltage across the SCR drops to zero, current will cease. However, if there is a substantial amount of inductance present, the thyristor will not turn off when reverse-biased and will resume conduction on the next half-cycle as soon as the forward-bias voltage is sufficient to maintain conduction, with a resulting loss of control. With resistive or slightly inductive loads the range of control is from 0 to 100 percent.

The disadvantage of the circuit shown in Fig. 4-3 is that, since there is not a common cathode connection, the gate circuits must be isolated. This disadvantage is overcome in the circuit of Fig. 4-5. There is a slight loss of efficiency, because an SCR and a diode are conducting at the same time, as well as an increase in the cost, but now that the cathodes are commoned, the

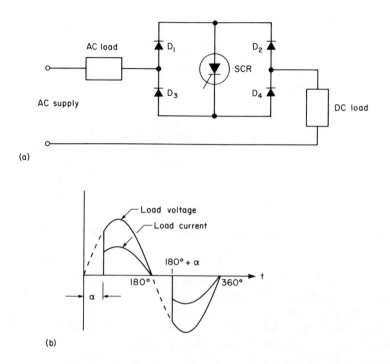

(a)

(b)

FIG. 4-4 Full-wave controller with the capability of phase controlling an ac or dc load. (a) basic circuit; (b) load voltage and current waveforms for an ac load.

gating circuitry does not have to be isolated, and the reverse voltages are blocked by the diodes. In all other respects the circuit functions in exactly the same way as that shown in Fig. 4-3.

The circuit of Fig. 4-6 permits on-load tap changing to be accomplished and is used both with phase control and pulse burst modulation. It is most commonly used for resistive heating loads and reduces the thermal shock, thus increasing the element life. When SCRs 3 and 4 are alternately fired, the load is effectively connected to the reduced voltage tap T1, and the heater output is held at a reduced level. If the full output of the heater is required, then SCRs 1 and 2 are alternately fired and SCRs 3 and 4 are biased off. This method of power control causes the least amount of distortion in the load waveform.

The circuit can also be used as a synchronous tap changer, in the following manner. SCR3 is gated on at the commencement of the positive half-cycle; after a firing delay angle of α deg, SCR1 is gated on and SCR3 turns off. SCR1 remains in conduction until it becomes reverse-biased and turns off (assuming

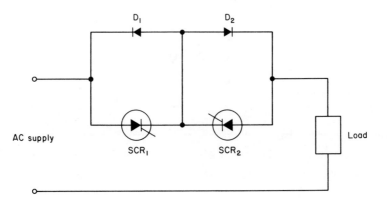

FIG. 4-5 Full-wave controller with common cathode connected SCRs.

a pure resistive load); then SCR4 is gated on, and α deg later SCR2 is turned on and SCR4 turns off, the process being repeated in synchronism with the ac source. The resulting load voltage waveform, illustrated in Fig. 4-6(b), is stepped with two voltage levels, the value of the rms output voltage being controlled by the firing delay angle. This method of power control produces less distortion than normal phase control techniques permit.

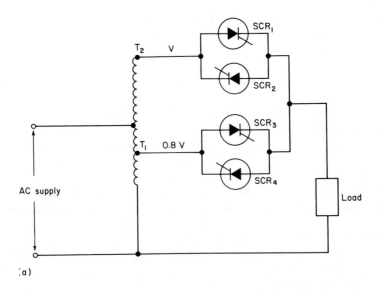

(a)

FIG. 4-6 Transformer tap changer. (a) basic circuit; (b) composite waveform when used as a synchronous tap changer with a resistive load.

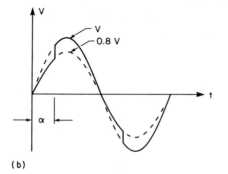

(b)

FIG. 4-6 (*cont.*)

The major difficulty with this configuration is that with inductive loads SCR1 will not be commutated off as it becomes reverse-biased. By gating SCR3 on, it will cease conduction; however, SCR4 cannot be turned on until the current has decreased to zero.

The circuits shown in Figs. 4-2 to 4-6 inclusive can be used with pulse burst modulation or phase control. However, they will, with the exception of Fig. 4-6, produce load waveforms that have high harmonic contents that increase as the firing delay angle increases up to 90 deg and reduce to zero when $\alpha = 180$ deg, as shown in Fig. 4-7.

4-3-2 Operation with Reactive Loads

When the circuits of Figs. 4-1 to 4-5 inclusive are operated with R-L loads, the following must be taken into consideration.

4-3-2-1 Thyristor Triggering

If the gate signal is a short duration pulse, a latching current may not be achieved because of the lag of the current with respect to the anode-cathode voltage. This may result in intermittent operation or no load current at all.

There are two basic solutions: First, connect a resistance in parallel with the load so that a current slightly in excess of the latching current will be drawn. Second, and more effective, is to apply a long-duration square pulse or a pulse train gate signal to ensure that conduction has been initiated.

Line synchronization must always be obtained for the gating circuit from the ac source, not from the voltage across the thyristor. If synchronization is attempted from the voltage across the thyristor, an unbalance in the positive and negative half-cycles of the load current will most likely occur with possible overloading of a thyristor and erratic control.

4-3-2-2   *di/dt*

The localized heating resulting from *di/dt* is not normally a serious problem, since the load inductance will limit the *di/dt*.

4-3-2-3   *dv/dt*

In Fig. 4-3 with a large delay angle and an inductive load when SCR1 turns off, SCR2 will be subjected to a high *dv/dt* as the source voltage is applied

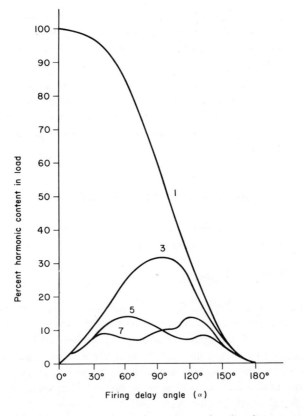

FIG. 4-7 Percent harmonic content of a resistive load voltage waveform vs. firing delay angle α.

across it. The effect of the dv/dt may trigger it on without a gate signal. This problem can be rectified by an R-C snubber network across each SCR.

4-3-2-4 Load Voltage Transients

When an inductive load is phase controlled, the switching off of the SCR may produce high-voltage transients ($-Ldi/dt$) of the order of several kilovolts. R-C snubbers across the SCRs will increase the turn-off time and reduce the di/dt and thus reduce the magnitude of the high-voltage transients; alternatively, a voltage-dependent resistor can be shunted across the inductive load.

4-4 FIRING CIRCUITS

There are a number of standard firing circuits that are used in phase-shift applications. Some of the more common circuits are dealt with in this section.

The basic block diagram of a phase-shift control system is shown in Fig. 4-8. The phase-shift control is synchronized to the ac source and generates the firing delay angle. The function of the trigger is to obtain a sufficiently strong pulse that will trigger a wide range of thyristors, without modifying the design of the circuit.

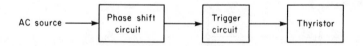

FIG. 4-8 Block diagram of a complete phase shift control of a thyristor.

4-4-1 The R-C Phase Shift and Extended R-C Phase-Shift Circuit

The simplest R-C phase-shift circuit is shown in Fig. 4-9. This depends upon the fact that the voltage drop across the resistance, V_R, leads the voltage drop across the capacitor, V_C or V_{OUT}, by 90 deg, the phasor sum of V_R and V_C at all times being equal to V_{ac}. The phase shift between V_{OUT} and V_{ac} is the firing delay angle α, as shown in Fig. 4-9(b). Increasing R will increase V_R and increase α, and reducing R will reduce V_R and α. This circuit theoretically permits α to be varied from 0 to 90 deg; however, in actual practice a variation from 10 to 80 deg is more realistic.

To increase the range of firing angle control the basic R-C phase shift

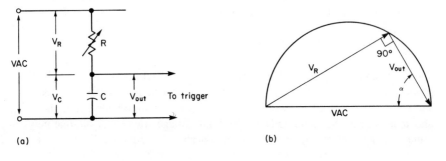

(a)   (b)

FIG. 4-9 Basic R-C phase shift. (a) basic circuit; (b) phasor diagram interrelating V_R, V_{OUT}, V_{AC} and α.

circuit is modified as shown in Fig. 4-10. This consists of a bridge circuit with two equal-value resistances $R1$ and $R2$ connected across the ac source. The other arm of the bridge consists of the variable resistance R and the capacitor C in series. The output voltage V_{OUT} is taken from points A and B. As can be seen from the phasor diagram, as R is increased to infinity α approaches 180 deg, and as R is decreased to zero, α decreases to 0 deg.

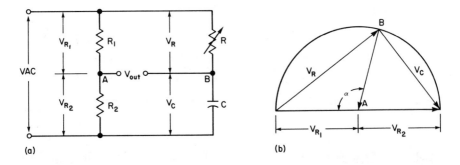

(a)   (b)

FIG. 4-10 Extended R-C phase shift control. (a) basic circuit; (b) phasor diagram.

The function of a triggering device is to rapidly transfer the stored energy of a capacitor to the gate of the thyristor. The most commonly used devices all have negative resistance characteristics, which at the point of breakdown act as a low-resistance switch, and transfer the stored energy. Typical devices are the unijunction transistor (UJT), the bidirectional diode thyristor (DIAC), the programmable unijunction transistor (PUT), and the silicon-controlled switch (SCS).

The more commonly used configurations, shown in Fig. 4-11, are satisfactory for half-wave or TRIAC control. In the case of an inverse-parallel SCR arrangement, firing control in both halves of the cycle can be obtained by modifying the UJT relaxation oscillator, as shown in Fig. 4-12.

The major disadvantage of the phase-shift controls illustrated in Figs. 4-11 and 4-12 is that relatively large variations of the variable resistance R are required to change the load voltage; in addition, the response is not linear. The transfer characteristic of the system can be improved; that is, the variation of load voltage can be made more proportional to the firing delay angle, by using a ramp and pedestal control.

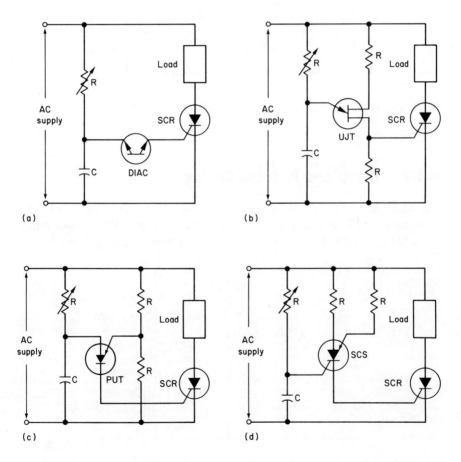

FIG. 4-11   R-C phase shift control circuits using as trigger devices (a) a DIAC, (b) a UJT, (c) a PUT, (d) an SCS.

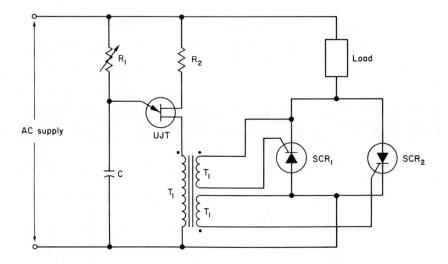

FIG. 4-12 Phase control firing circuit for inverse parallel SCRs.

4-4-2   Ramp and Pedestal Control

The basic circuit of a ramp and pedestal control is shown in Fig. 4-13. The capacitor C will charge very rapidly through the diode D to V_C, and this estab-lishes the pedestal voltage. The capacitor will continue charging exponentially

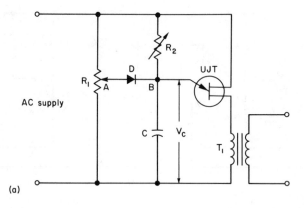

(a)

FIG. 4-13 Ramp and pedestal control. (a) basic circuit; (b) ramp and pedestal voltages.

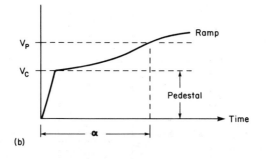

FIG. 4-13 (b) ramp and pedestal voltages.

at a rate determined by the time constant $R2C$. When the potential at point B rises above the potential at point A, the diode becomes reverse-biased, and the capacitor continues charging along the ramp until it reaches the triggering voltage V_p of the UJT. The ramp is not linear, since it is supplied from the sinusoidal ac source. The slope of the ramp is determined by the adjustment of $R2$; the firing delay angle is varied by varying $R1$ and thus the voltage V_C across the capacitor. Varying $R1$ to increase the pedestal voltage decreases α, or vice versa.

4-4-3   Cosine Modified Ramp and Pedestal Control

Further improvements in the linearity of the ramp can be made by using the circuit of Fig. 4-14. In this circuit the capacitor C is supplied directly from the ac line, thus adding a cosine function to the nonlinear ramp, with the result that the ramp becomes nearly linear and the transfer characteristic is also nearly linear, and the gain is constant.

4-5   THREE-PHASE APPLICATIONS

Most of the single-phase circuits described in Sec. 4-3-1 can be applied in three-phase applications, some of which are illustrated in Fig. 4-15. The firing circuit requirements are that the thyristors be fired in pairs and that the gate signals be spaced 120 deg apart and be of sufficient duration that a latching current is achieved. For example, in Fig. 4-15(c) the firing sequence would be SCRs 1, 4, 5; 4, 5, 2; 5, 2, 3; 2, 3, 6; 3, 6, 1; and 6, 1, 4.

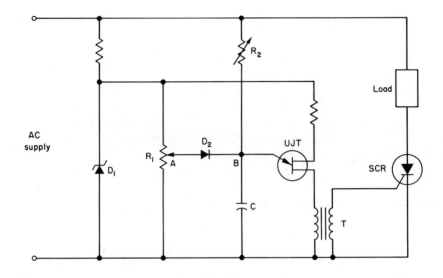

FIG. 4-14 Cosine modified ramp and pedestal control.

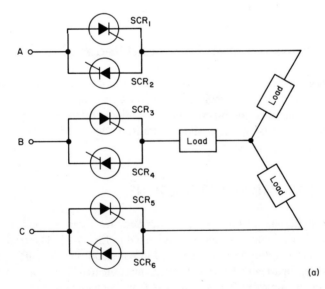

FIG. 4-15 Three-phase thyristor circuits controlling ac loads.

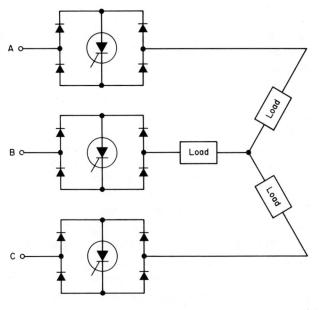

(b)

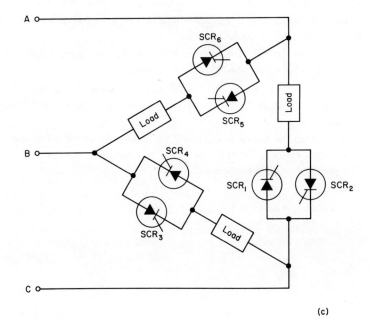

(c)

FIG. 4-15 *(cont.)*

4-6   SUMMARY

The circuits described in this chapter are most commonly used for both resistive and inductive applications. The pulse burst modulation technique can be used in single- and three-phase applications, provided that there is sufficient thermal or mechanical inertia in the load (heating or motor control). However, the most popular and simplest method used is phase-shift control, but at low-power levels a significant harmonic content is introduced into the load waveform; this may be minimized by using the synchronous tap-changer concept. Also, significant amounts of radio frequency interference may be generated, which must be filtered out.

REVIEW QUESTIONS

1. Discuss the principle of pulse burst modulation and its range of control.
2. Discuss with the aid of schematics the advantages and disadvantages of the more common methods of single-phase ac phase-control circuits.
3. Explain what is meant by the term *synchronous tap changing*.
4. What are the requirements for successful thyristor triggering in ac phase control, and how may the effects of dv/dt and load transients be minimized?
5. Explain with the aid of a schematic and a phasor diagram the principle of the R-C phase-shift control circuit.
6. Explain with the aid of a schematic and a phasor diagram the principle of the extended R-C phase-shift control circuit.
7. Explain the principle of phase-shift control using the ramp and pedestal technique. What is its major advantage compared to the R-C phase-shift circuits?
8. Explain with the aid of a schematic the cosine modified ramp and pedestal phase-shift control. What is the major advantage as compared to the ramp and pedestal control technique?

5 Thyristor Phase-Controlled Converters

5-1   INTRODUCTION

Phase-controlled converters are line-commutated ac-dc converters that produce a variable dc output voltage, whose magnitude is varied by phase control that, is, by controlling the duration of the conduction period by varying the point at which a gate signal is applied to the rectifying device. This principle has in the past been applied to thyratrons, mercury arc rectifiers, such as ignitrons and excitrons, and now to thyristors.

In most converters the power flow is from the ac source to the dc load, which is the rectification process. However, some converter configurations permit control so that power may flow from the dc load under regenerative conditions back into the ac source. This method of transferring energy from the dc load to the ac source is known as *synchronous inversion*, as distinct from inversion of dc to variable frequency ac by means of a dc link converter, utilizing forced commutation techniques.

Converters range from the simplest configuration of the half-wave single-phase rectifier, which is seldom used in power electronic applications because of the high ripple voltage content in its output, to the bridge circuit and the midpoint circuit.

In addition, converters are also classified as to whether the ac voltage source is single-phase or three-phase. Further increasing the number of phases by suitable transformer connections increases the ripple frequency of the output dc voltage, thus reducing the filtering requirement. The ratio of the ripple frequency to the ac source frequency is known as the *pulse number*.

If all the active devices in the converter are diodes, the converter is called an uncontrolled rectifier, and its dc output is dependent upon the amplitude of the input ac voltage and the converter configuration. Replacement of half of the active devices by thyristors produces a half-controlled converter or semiconverter. Semiconverters permit the average value of the dc output voltage to be varied by phase angle control of the thyristors, but the converter is still a rectifier; that is, power flow is from the ac source to the dc load. Converters of this type are called *one-quadrant* converters, since the polarity of the dc voltage and the current direction cannot change.

If all the active devices in a converter are thyristors, the converter is classified as a full converter or two-quadrant converter. The resultant dc current is unidirentional, but the dc voltage may have either polarity. With one polarity the power flow is from the ac source to the dc load, that is, rectification, and with a reversal of the dc voltage by the load, the power flow is from the dc source to the ac supply, the process known as *synchronous inversion*.

In applications such as mine hoists, the converter must have the capability of providing both polarities of dc voltage and dc current; converters operating in this mode are known as *four-quadrant* converters. Four-quadrant operation is obtained by reversing the output of the two-quadrant converter by a reversing switch, or by connecting two two-quadrant converters "back-to-back" forming a dual converter.

Phase-controlled converters in the above configurations provide dc power to such diverse applications as dc motor control in rolling mills, paper mills, and mine hoists; battery charging; electrochemical processes; and high-voltage dc transmission systems.

5-2   THE BASIC PRINCIPLE OF PHASE CONTROL

The concepts and terminology used in phase control can be best described as a result of considering a half-wave phase-controlled thyristor with resistive and inductive loading, as illustrated in Fig. 5-1.

During the negative half-cycle of the supply voltage, the thyristor blocks the flow of load current, and no voltage is applied to the load.

During the positive half-cycle of the supply voltage, the thyristor is forward-biased and will conduct if gated on. If the thyristor is turned on at time t_1, load current will flow and the supply voltage minus the device drop of approximately one volt is applied to the load [Fig. 5-1(b) and (c)]. In the case of a pure resistive load when the thyristor becomes reverse-biased at time t_2, load current will cease, and no voltage will be applied to the load until the

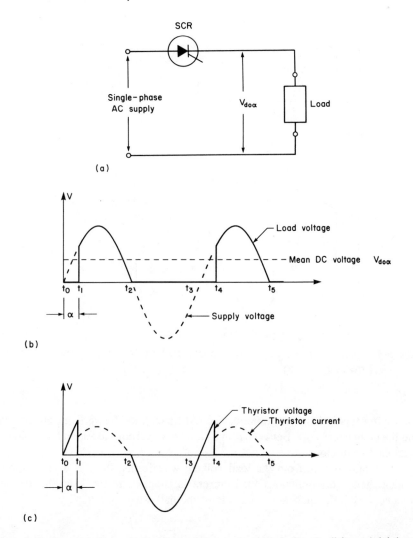

FIG. 5-1   Single-phase, half-wave phase control. (a) Circuit; (b) and (c) load
voltage, thyristor voltage, and current waveforms, resistive loading.

thyristor is forward-biased and gated on at time t_4. The amplitude of the mean
dc voltage $V_{do\alpha}$ is controlled by the firing delay angle α.

In the case of an inductive load [Fig. 5-1(d) and (e)], if the thyristor is
turned on at time t_1, load current will flow, and the supply voltage minus the
device drop is applied to the load. However, at time t_2, when the thyristor is
reverse-biased, the stored magnetic energy in the load is returned and sustains

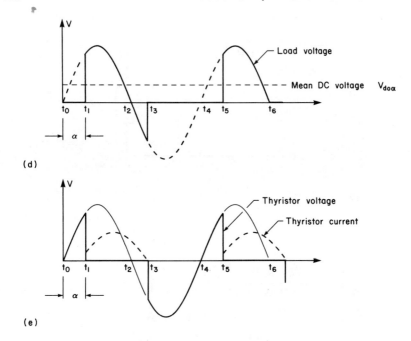

FIG. 5-1 (*cont.*) (d) and (e) load voltage, thyristor and current waveforms, inductive loading.

a decaying forward current in the device. At time t_3, the forward current ceases and the device turns off. Between t_3 and t_5, no load current flows, and no voltage is applied to the load.

As can be seen from the load voltage waveforms, there is a high ripple content, and a discontinuous load current is supplied to the load. In normal practice very little application is made of the half-wave circuit.

5-3   TWO-QUADRANT CONVERTERS

For two-quadrant operation, that is, unidirectional dc current flow and both polarities of dc voltage, it is necessary to have thyristors in all positions.

In the remaining discussions for simplicity it is assumed that:

1. The voltage drop across the device is negligible when it is conducting.
2. There is no leakage current when the device is blocking.
3. Turn-on and turn-off occur instantly.

4. The output dc terminals are connected to an infinite inductance (i.e., an ideal filter), thus producing a ripple-free constant amplitude current.

In general two-quadrant converters can be classified as:

1. Midpoint converters which are supplied from tapped transformers.

2. Bridge converters which may be supplied directly from the ac source.

5-3-1   Two-pulse Midpoint Converter

Figure 5-2 shows the basic arrangement of a single-phase, two-pulse midpoint converter with source voltage and current, load voltage and current, and thyristor voltage and current waveforms, for various firing delay angles α. Consider Fig. 5-2(b), where $\alpha = 0$ deg.

During the positive half-cycle of the supply voltage, SCR1 is forward-biased, the transformer secondary S_1 carries the load current, the load voltage is V_{S1}, and SCR2 is reverse-biased. During the negative half-cycle SCR2 is forward-biased, the transformer secondary S_2 carries the load current, and the load voltage is V_{S2}, and SCR1 is reverse-biased.

Application of a gate signal with a zero firing delay angle results in the converter acting as an uncontrolled rectifier. The load voltage waveform consists of a dc component V_{do} and a superimposed ac ripple with a frequency of twice that of the supply, accounting for the name of two-pulse converter.

The total dc output current I_d is supplied in turn by SCR1 and SCR2, and is assumed to be square; i.e., it consists of a fundamental and harmonics. The load current I_d flowing in the secondary is in phase with the secondary voltage, and the converter draws a load current from the supply at unity power factor.

Observe Fig. 5-2(c), where $\alpha = 45$ deg. Increasing the firing delay angle

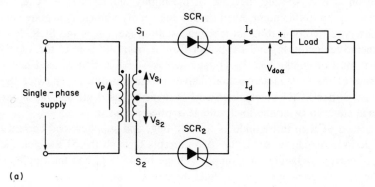

(a)

FIG. 5-2 Two-pulse midpoint converter. (a) Basic circuit.

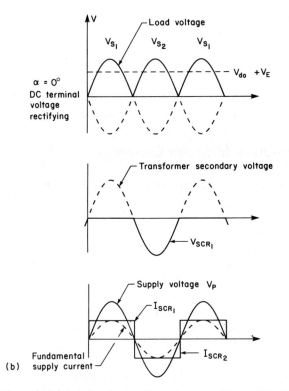

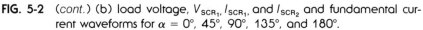

FIG. 5-2 (*cont.*) (b) load voltage, V_{SCR_1}, I_{SCR_1}, and I_{SCR_2} and fundamental current waveforms for $\alpha = 0°$, $45°$, $90°$, $135°$, and $180°$.

α produces a decrease in the average dc voltage $V_{do\alpha}$. If the load is highly inductive, the load current I_d will remain approximately constant. If SCR1 is gated on at $\alpha = 45$ deg, because of the highly inductive load SCR2 will be conducting up to this point, even though the supply voltage is negative and can be observed by connecting an isolated oscilloscope across the SCR. This is a regenerative condition, and the power flow is from the load to the supply. Up to the point of conduction the voltage across SCR1 is positive and is equal to the sum of the voltage across both halves of the transformer or twice the load voltage. When SCR1 conducts, SCR2 is reverse-biased off and has a negative voltage equal to twice the load voltage applied across it.

Each SCR in turn conducts for 180 deg, and supplies load current I_d for this period. As can be seen, the load currents through the SCRs, and thus the supply current is lagging the supply voltage by 45 deg, and the ripple content of the dc voltage supplied to the load increases.

If the load is resistive, then as each SCR in turn is reverse-biased, load current will cease until the next gate signal is received, and the current supplied

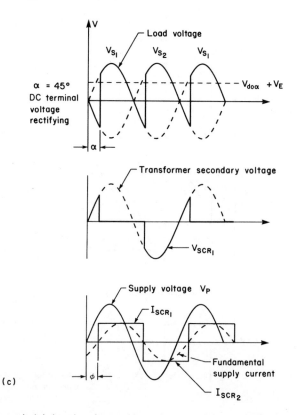

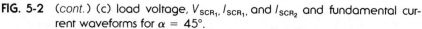

FIG. 5-2 *(cont.)* (c) load voltage, V_{SCR_1}, I_{SCR_1}, and I_{SCR_2} and fundamental current waveforms for $\alpha = 45°$.

to the load will be discontinuous.

Observe Fig. 5-2(d), where $\alpha = 90$ deg. When the firing delay angle is increased to 90 deg, the average dc voltage $V_{do\alpha}$ is zero, and the load voltage consists entirely of the ac ripple components. The SCRs, if an inductive load is assumed, remain in conduction for 180 deg, and the SCR currents are now lagging the appropriate anode-cathode voltage by 90 deg. As a result, the supply current lags the supply voltage by 90 deg, and there is no transfer of power from the ac source to the dc load.

In summary, as the firing delay angle is increased, the power factor of the converter is worsened, and the power transferred to the dc load reduces to zero when $\alpha = 90$ deg $(V_p I_p \cos \phi = 0)$.

In Fig. 5-2(e), where $\alpha = 135$ deg, when the firing delay is increased beyond 90 deg load current can flow only if the load itself presents a negative voltage, which will occur in the case, for example, of a dc motor under overhauling load conditions.

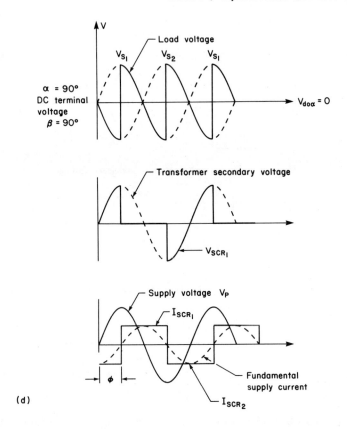

FIG. 5-2 (cont.) (d) load voltage, V_{SCR_1}, I_{SCR_1}, and I_{SCR_2} and fundamental current waveforms for $\alpha = 90°$.

When $\alpha = 135$ deg, the average dc voltage $V_{do\alpha}$ is negative. The load current flows in each SCR for 180 deg in its original direction, but the voltage has reversed polarity; the power flow is from the dc load to the ac source, and the converter is acting as a line-commutated inverter. The process is known as *synchronous inversion*.

The SCR currents are lagging by 135 deg, and as a result the supply current lags the source voltage by 135 deg.

In Fig. 5-2(d), where $\alpha = 180$ deg, when the firing delay angle is increased to 180 deg, the mean dc voltage $V_{do\alpha}$ has reached its maximum negative value, and the ac ripple content has decreased from its maximum at $\alpha = 90$ deg to a minimum at $\alpha = 180$ deg. The SCRs remain in conduction for 180 deg, and the SCR currents and the resulting ac supply current lag the source voltage by 180 deg.

In actual practice this situation cannot be achieved, since the period of

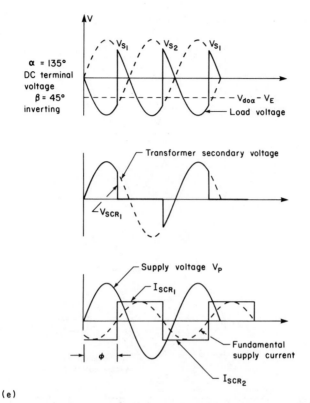

$\alpha = 135°$
DC terminal
voltage
$\beta = 45°$
inverting

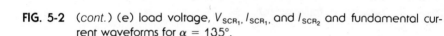

(e)

FIG. 5-2 (cont.) (e) load voltage, V_{SCR_1}, I_{SCR_1}, and I_{SCR_2} and fundamental current waveforms for $\alpha = 135°$.

reverse bias of the thyristors is continually reducing as α approaches 180 deg. Sufficient time must be allowed for the thyristors to turn off and regain forward blocking capability before forward voltage is reapplied; otherwise, commutation failure will occur. Normally for 60 Hz systems the maximum delay angle is limited to about 160 to 175 deg.

When operating in the inverting mode, the inversion region is specified in terms of the inverter advance angle β, where $\beta = 180° - \alpha$.

In summary, the following should be noted with respect to the operation and performance of a two-pulse midpoint converter:

1. For firing delay angles $0° < \alpha < 90°$, the converter is operating in the rectifying mode; for $90° < \alpha < 180°$ the operation is in the inversion mode.

2. In the rectifying mode the net transfer of power is from the ac source

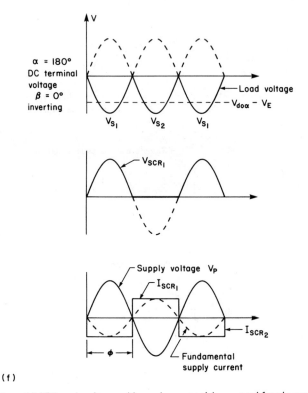

(f)

FIG. 5-2 *(cont.)* (f) load voltage, V_{SCR_1}, I_{SCR_1}, and I_{SCR_2} and fundamental current waveforms for $\alpha = 180°$.

to the dc load. In inversion, provided that there is a negative source of voltage at the dc load, the net transfer of power is from the dc load to the ac source.

3. As the firing delay angle is increased, there is a corresponding lag of the ac supply current with respect to the source voltage.

4. The mean dc voltage $V_{do\alpha}$ decreases from a positive maximum to zero at $\alpha = 90°$ to a negative maximum at $\alpha = 180°$, while the ac ripple content increases from a minimum at $\alpha = 0°$ to a maximum at $\alpha = 90°$ and then decreases to a minimum at $\alpha = 180°$.

5. With a sufficiently inductive load the thyristors remain in conduction for 180 deg, and the load current I_d is continuous. In the case of a pure resistive load, conduction ceases when the conducting thyristor becomes reverse-biased, and the load current is discontinuous.

6. The thyristors must be rated to withstand at least twice the rated peak load voltage.

Midpoint converters are used in applications where electrical isolation is required, and, as will be shown in the next section, where it is desirable to reduce the ac ripple content by increasing the number of phases. The two-pulse midpoint converter has been used to illustrate the concept, but operation from single-phase sources is limited by the power capability of the source, and most industrial applications utilize polyphase converters.

Midpoint converters, while they use only half the number of thyristors required in a bridge converter, find economies in the cost and simplicity of the control section, which is offset to some extent by the increased cost of the higher voltage rating of the thyristors.

5-3-2 Two-pulse Bridge Converter

The single-phase converter shown in Fig. 5-3 is an alternative two-pulse converter which eliminates the requirement for an input transformer.

In this circuit diagonally opposite pairs of thyristors conduct and commutate together. The control is identical to that of the two-pulse midpoint converter; i.e., it controls the mean output dc voltage $V_{do\alpha}$ from a maximum positive to zero, the rectifying mode, and if there is a negative voltage source at the load, from zero to a maximum negative in the inversion mode.

The major differences of the bridge converter as compared to the midpoint converter are:

1. There are always two thyristors in conduction at the same time, resulting in a device voltage drop double that of the two-pulse midpoint converter.

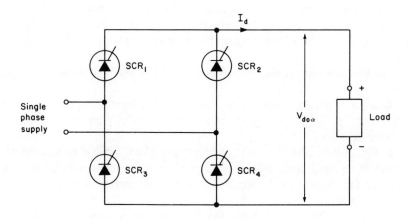

FIG. 5-3 Two-pulse bridge converter.

2. The peak applied voltage is also the peak voltage applied to each thyristor.

3. The control circuitry is of necessity slightly more complex.

5-3-3   Three-pulse Midpoint Converter

In order to reduce the ac ripple content of the output dc voltage, and to minimize the need for smoothing, as well as to increase the power output capability of a converter, a three-phase version of the midpoint converter was developed.

A three-phase ac supply combined with suitable transformer connections permits an increase in the pulse number, or an increase in the number of dc voltage segments produced for each cycle of the ac supply voltage. The higher the pulse number, the smoother is the output dc voltage.

The simplest version is the three-pulse midpoint converter, illustrated in Fig. 5-4.

The transformer is star-zigzag connected. The zigzag connection requires two separate identical secondary windings on each phase, and for the same output voltage requires more copper than a single winding secondary. The function of the zigzag connection is to prevent dc magnetization of the core by the dc component of the ac supply circuit, by permitting equal and opposite currents to flow in each half of the secondary winding.

From Fig. 5-4(b) it can be seen that each thyristor is fired at the point when it becomes forward-biased, that is, $\alpha = 0°$, or 30° after the phase voltage crosses the zero axis, and as a result the ac input voltage with the greatest instantaneous value is applied to the dc load terminals, and the mean dc output voltage V_{do} is at its positive maximum. The conduction period of each thyristor is 120 deg, and it blocks reverse voltage for 240 deg, the maximum blocking voltage being equal to the line-to-line voltage of the ac source. The frequency of the ac ripple content of the dc output voltage is $3f$, where f is the ac source frequency.

The thyristor current also has a duration of 120 deg, and its amplitude is $I_d/3$.

When $\alpha = 0°$, the supply voltage and the fundamental component of the supply current are in phase, and the converter appears as a unity power factor load to the ac source.

In Fig. 5-4(c), when $\alpha = 45°$, the thyristors block forward voltage for 45 deg from the natural commutation point, the mean dc output voltage $V_{do\alpha}$ is reduced, and the fundamental component of the supply current lags the ac source voltage by 45 deg.

When $\alpha = 90°$ [Fig. 5-4(d)], the thyristors block forward and reverse voltages for equal periods, and $V_{do\alpha}$ is zero; the angle of phase lag of the ac

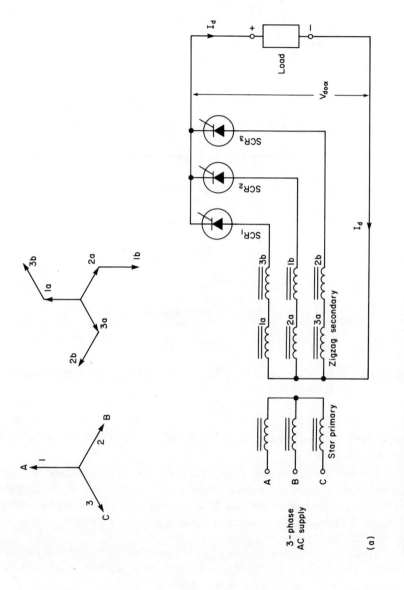

FIG. 5-4 Three-pulse midpoint converter. (a) Basic circuit.

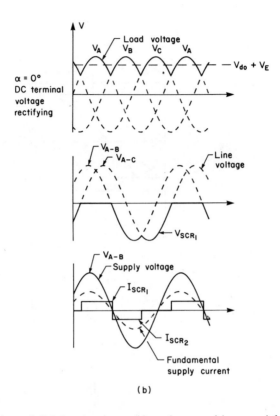

FIG. 5-4 (*cont.*) (b) load voltage, V_{SCR_1}, I_{SCR_1}, and I_{SCR_2} and fundamental current waveforms for $\alpha = 0°$.

supply current with respect to the ac supply is the same as the firing delay angle, i.e., $\phi = \alpha = 90°$.

For firing delay angles greater than 90 deg, it is necessary to have a counter-voltage source at the load. When $\alpha = 135°$ [Fig. 5-4(e)], the thyristors are blocking voltage in the forward direction, and $V_{do\alpha}$ is becoming increasingly negative; the fundamental component of the supply current lags the supply voltage by 135 deg.

When $\alpha \simeq 180°$, the thyristors block forward voltage for almost the whole 180 deg, allowing only sufficient time for turn-off after a very short conduction period. $V_{do\alpha}$ is now at its negative maximum.

For $0° < \alpha < 90°$ the converter is rectifying, and for $90° < \alpha < 180°$ it is operating in the inversion mode as a synchronous inverter.

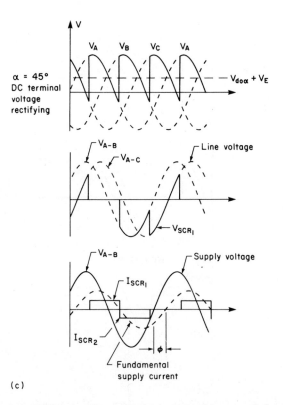

(c)

FIG. 5-4 (cont.) (c) load voltage, V_{SCR_1}, I_{SCR_1}, and I_{SCR_2} and fundamental current waveforms for $\alpha = 45°$.

5-3-4 Six-pulse Midpoint Converters

To further reduce the ac ripple content of the mean dc output voltage, the pulse number can be increased to six by using a six-pulse midpoint converter or a six-pulse midpoint converter with an interphase reactor.

5-3-4-1 Six-pulse Midpoint Converter

In its simplest configuration it is supplied with a three-phase to six-phase transformer. However, it has a serious disadvantage; namely, each thyristor conducts only for 60 deg, resulting in both the thyristors and the transformer secondaries being poorly utilized because of the high ratio of rms to average current. As a result, this configuration is rarely used. (See Fig. 5-5.)

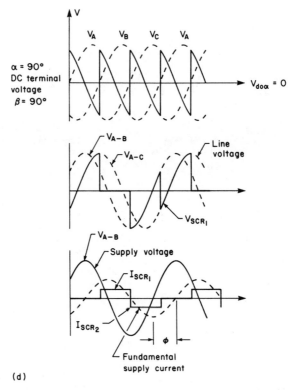

(d)

FIG. 5-4 (*cont.*) (d) load voltage, V_{SCR_1}, I_{SCR_1}, and I_{SCR_2} and fundamental current waveforms for $\alpha = 90°$.

5-3-4-2   Six-pulse Midpoint Converter with Interphase Reactor

This converter arrangement combines the outputs of two three-pulse midpoint converters operating in parallel through an interphase reactor. Each converter operates independently of the other, the load current being shared equally between the converters, with the reactor absorbing the difference in the instantaneous voltages. The ac ripple content of the outputs of the individual converters is $3f$, however, because of the phase difference between the ac ripple contents of the two converters, the mean output dc voltage at the center tap of the interphase reactor has an ac ripple content of $6f$. As a result, for each cycle of the primary supply there are six segments of dc voltage in the mean output dc voltage, and unlike the previous configuration each thyristor conducts for 120 deg, resulting in an improved transformer utilization.

The operation depends upon the interphase reactor carrying sufficient dc current to maintain an adequate flux level in its core. If on light loads the core

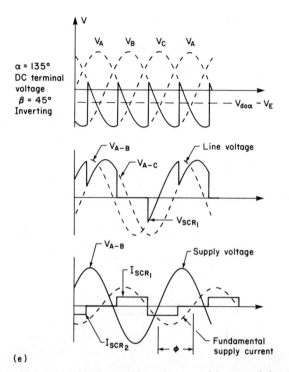

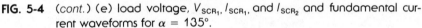

(e)

FIG. 5-4 (cont.) (e) load voltage, V_{SCR_1}, I_{SCR_1}, and I_{SCR_2} and fundamental current waveforms for $\alpha = 135°$.

flux drops, the converter will operate as a normal six-pulse midpoint converter with the thyristors conducting for 60 deg, producing a discontinuous output with a significant drop in the mean output dc voltage. This problem can be minimized by connecting a bleeder resistance across the output terminals, ensuring a minimum dc current through the interphase reactor.

The basic circuit and the associated waveforms are shown in Fig. 5-6 for firing delay angles of $\alpha = 45°$ and $135°$. It should be noted that the transformer primary current no longer has a dc component present and as a result more closely approximates a sine wave than has been the case in the previously considered converter configurations.

5-3-5 Higher Pulse Number Midpoint Converters

The pulse number can be increased to twelve by simply connecting two six-pulse midpoint converters in parallel with each other through a third interphase reactor, to form a twelve-pulse midpoint coverter. (See Fig. 5-7.) The main

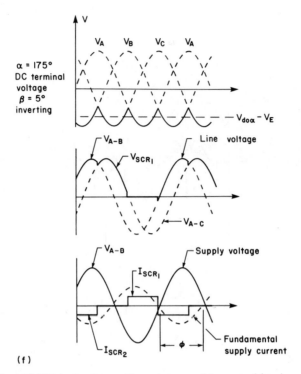

FIG. 5-4   (*cont.*) (f) load voltage, V_{SCR_1}, I_{SCR_1}, and I_{SCR_2} and fundamental current waveforms for $\alpha = 175°$.

reasons are that the mean output dc voltage is smoother with a $12f$ ac ripple content, the thyristors still conduct for 120 deg, and the transformer primary current is even closer to being sinusoidal.

5-4   SIX-PULSE BRIDGE CONVERTERS

The six-pulse bridge converter consists of two three-pulse midpoint converters connected in series, as shown in Fig. 5-8(a). The currents that flow in the neutral are equal and opposite, and as a result the neutral line becomes redundant. The converter configuration shown in Fig. 5-8(b) is the result of eliminating the neutral. The load voltage $V_{do\alpha}$ is the sum voltage of the half-wave outputs of the two series-connected three-pulse converters. The upper group with a common cathode connection, when $\alpha = 0°$ [Fig. 5-8 (c)], will result in the thyristor with the most positive anode conducting and producing a positive potential at the cathode busbar. Similarly, the lower group has a common anode connection, and when $\alpha = 0°$ the thyristor with the most negative cathode will be in con-

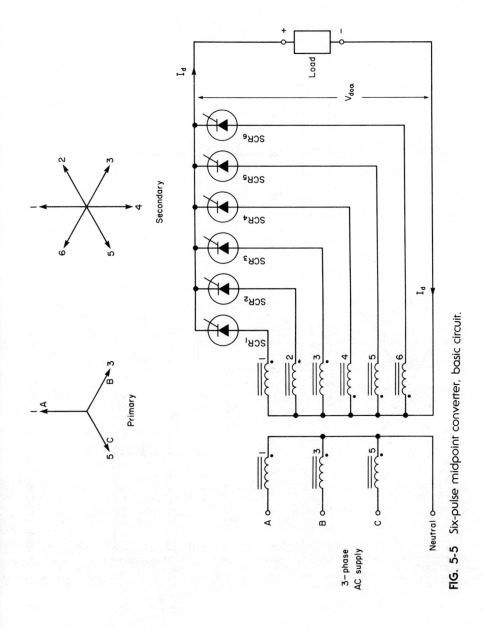

FIG. 5-5 Six-pulse midpoint converter, basic circuit.

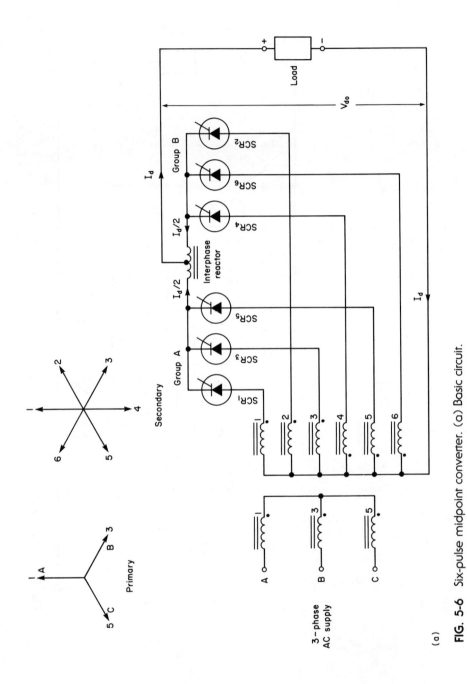

FIG. 5-6 Six-pulse midpoint converter. (a) Basic circuit.

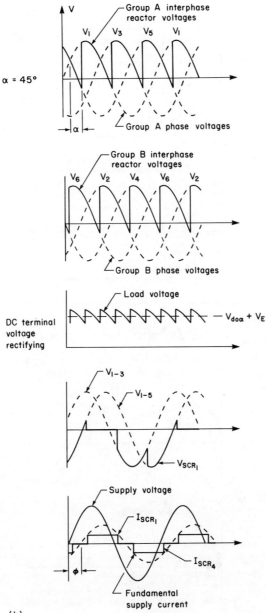

(b)

FIG. 5-6 (*cont.*) (b) interphase reactor voltages, load voltage, V_{SCR_1}, I_{SCR_1}, and I_{SCR_4} and fundamental current waveforms for $\alpha = 45°$.

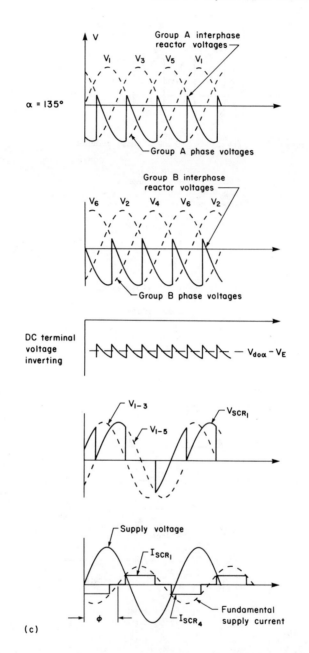

FIG. 5-6 (cont.) (c) interphase reactor voltages, load voltage, V_{SCR_1}, I_{SCR_1}, and I_{SCR} and fundamental current waveforms for $\alpha = 135°$.

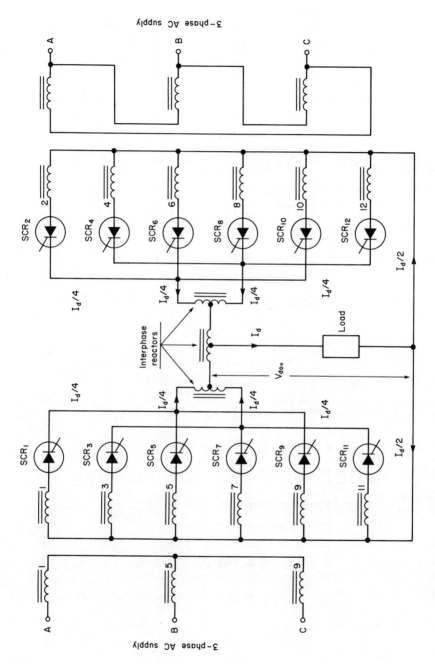

FIG. 5-7 Twelve-pulse midpoint converter.

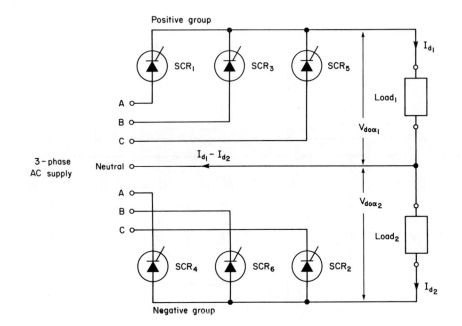

(a)

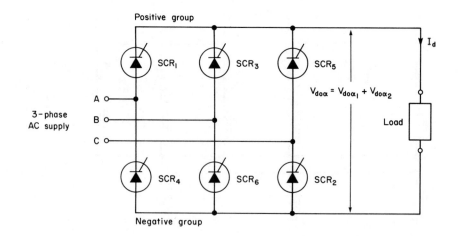

(b)

FIG. 5-8 Six-pulse bridge converter. (a) Basic concept; (b) basic circuit.

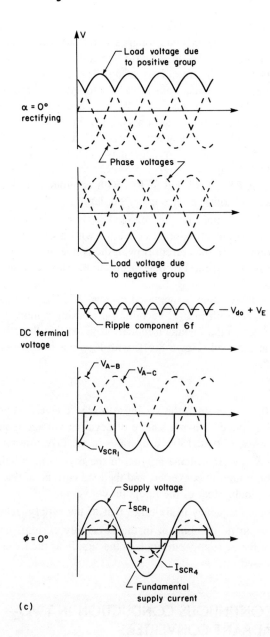

FIG. 5-8 (*cont.*) (c) load voltages due to positive and negative groups, load voltage, V_{SCR_1}, I_{SCR_1}, and I_{SCR_4}, supply and fundamental current waveforms for $\alpha = 0°$.

duction, making the common anode busbar negative. The output voltage V_{do} is the algebraic sum of the instantaneous voltages at the cathode and anode busbars, and consists of segments of the three-phase line-to-line voltages and will have an ac ripple component with a frequency of $6f$. The magnitude of V_{do} is twice that of the individual three-pulse converters. Each thyristor carries the dc load current I_d for 120 deg and blocks for 240 deg, and there is no dc current component in the ac line current. There must be two thyristors in conduction at the same time, although the conduction periods of the thyristors in the upper and lower groups do not occur simultaneously. The firing order for a supply phase sequence ABC would be SCR1 and SCR2, SCR2 and SCR3, SCR3 and SCR4, SCR4 and SCR5, SCR5 and SCR6, with commutation occurring alternately in the upper and lower groups every 60 deg.

When $\alpha = 45°$, the mean output voltage $V_{do\alpha}$ is reduced [Fig. 5-8(d)] and is twice that of a three-pulse midpoint converter with a zero dc component in the ac supply current, since the current flow in the converter consists of positive and negative blocks of current 120 deg in duration. The ac ripple component has a frequency of $6f$.

As the firing delay angle is increased to $\alpha = 90°$, $V_{do\alpha}$ decreases to zero. Further increases in α result in $V_{do\alpha}$ becoming increasingly more negative, and the converter operates in the inversion mode, provided that there is a source of negative voltage at the load [Fig. 5-8(e)], with $V_{do\alpha}$ reaching a negative maximum at $\alpha \simeq 180°$.

In summary,

1. When $0° < \alpha < 90°$, the six-pulse bridge converter operates in the rectifying mode. When a source of negative voltage is present at the load for $90° < \alpha < 180°$, it operates in the inverting mode.

2. The peak reverse voltage applied to the thyristors is half that of the six-pulse midpoint converter, and the load current carried by the thyristors is double that of the midpoint converter.

3. The ripple frequency is six times that of the supply frequency.

4. There is no dc component in the ac supply current, and the phase angle of the ac supply current lags the supply as the firing delay angle is increased.

5-5 DISCONTINUOUS CONDUCTION IN TWO-QUADRANT CONVERTERS

In all the considerations to date, it has been assumed that the load has been highly inductive and as a result the dc load current I_d is continuous. In actual practice this is not always the case; the effects on the load voltage waveform

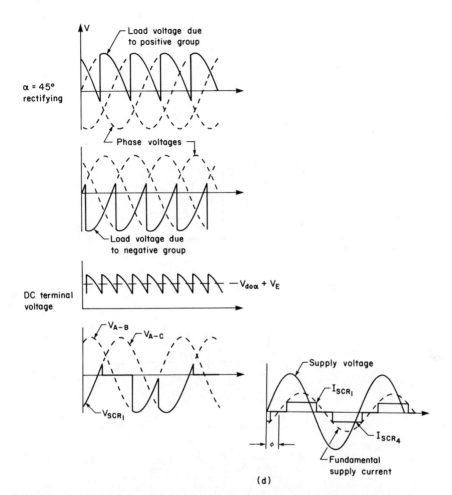

FIG. 5-8 (cont.) (d) load voltages due to positive and negative groups, load voltage, V_{SCR_1}, I_{SCR_1}, and I_{SCR_4}, supply and fundamental current waveforms for $\alpha = 45°$.

must be considered for passive R-L loads and back emf loads with various load current conditions.

5-5-1 Passive R-L Loads

The load Q factor can vary from being infinite, i.e., pure inductance, to finite and zero, i.e., pure resistance. In our considerations to date Q has been assumed to be infinite and the load current I_d has been considered to be constant in

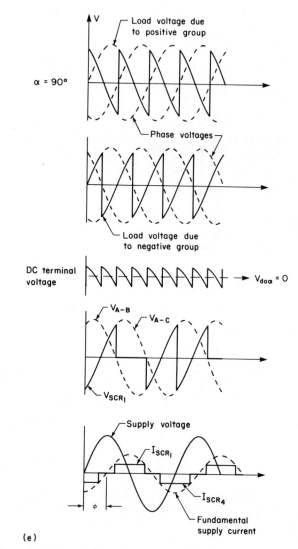

(e)

FIG. 5-8 (*cont.*) (e) load voltages due to positive and negative groups, load voltage, V_{SCR_1}, I_{SCR_1}, and I_{SCR_4}, supply and fundamental current waveforms for $\alpha = 90°$.

amplitude. This assumes that ac ripple current components in the dc circuit have been negligible. In a practical converter this is not so. Provided that the peak amplitude of the ac ripple current is less than the dc component, the resultant current has always been finite, and the converter has remained continuously conducting. If however, the negative peak amplitude of the ac ripple exceeds the amplitude of the dc component, the resultant current supplied to the load

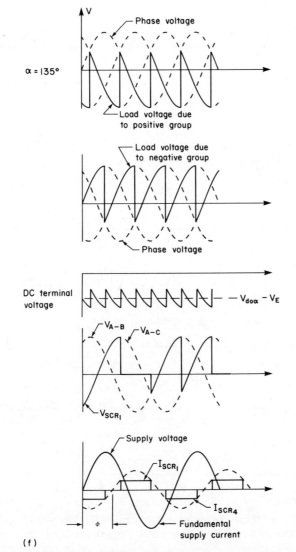

FIG. 5-8 (cont.) (f) load voltages due to positive and negative groups, load voltage, V_{SCR_1}, I_{SCR_1}, and I_{SCR_4}, supply and fundamental current waveforms for $\alpha = 135°$.

will be discontinuous because of the unidirectional properties of the converter, and the load voltage waveform will be modified.

The continuity of the load current waveform depends upon the load Q factor and the firing delay angle. Figure 5-9 illustrates the effects of varying load Q factors and firing delay angles for a six-pulse bridge converter.

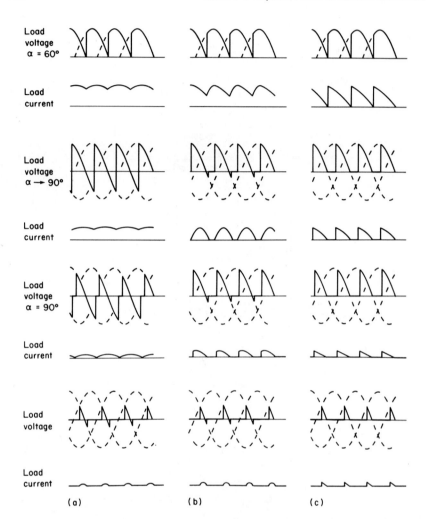

FIG. 5-9 Six-pulse bridge converter with a passive R-L load. (a) $Q \to \infty$ (infinite inductance); (b) $Q \simeq 3$ (normal R-L load); (c) $Q = 0$ (pure resistance).

When $Q \to \infty$, i.e., when L/R is very large, the load current remains continuous up to $\alpha \simeq 90°$, and the load voltage $V_{do\alpha}$ follows the ac supply voltage. When $\alpha = 90°$, the load current I_d will be zero or discontinuous for a very short period, and as a result there will be neither voltage nor current at the load terminals. Increasing α beyond 90° results in the load current becoming zero for greater periods, and $V_{do\alpha}$ approaches more closely to zero. As α approaches 120 deg, conduction ceases, and $V_{do\alpha}$ is zero. Refer to Fig. 5-9(a).

When the load Q factor is finite, the current becomes discontinuous for $\alpha < 90°$; as α is increased to 120 deg, the stored energy in the inductance is insufficient to maintain conduction, and the load current and voltage will be zero when $\alpha = 120°$ [see Fig. 5-9(b)].

For a pure resistive load, $Q = 0$, the load current becomes discontinuous at $\alpha = 60°$, and the load current and voltage will be zero at $\alpha = 120°$ [see Fig. 5-9(c)].

In summary, as the load Q factor decreases, a large variation in the firing delay angle results with a relatively small change in the mean load voltage $V_{do\alpha}$, and the load current becomes discontinuous for $60° < \alpha < 90°$, since for firing delay angles less than 60 deg, with a passive load the instantaneous value of the mean load voltage is positive.

5-5-2 Active Loads

If the load is capable of voltage storing, e.g., a capacitor or battery, or is a source of negative voltage such as the back emf of a dc motor, the load current will become discontinuous if it is reduced sufficiently, as would occur in the case of a dc motor with a fluctuating load.

Once again with reference to the case of a six-pulse bridge converter, it is possible for discontinuous conduction to occur in either the rectifying or inverting modes, if the load current is low enough. (See Fig. 5-10.)

Maintaining the firing delay angle α constant under conditions of varying load current results in the load voltage waveform following the ac supply voltage under continuous load current conditions. Under discontinuous load current conditions the load voltage rises to that of the "back emf," with the result that the mean load voltage has changed without a corresponding change in the firing delay angle. To maintain a constant mean load voltage between a light load with discontinuous conduction and full load with continuous conduction requires that the firing delay angle be advanced by the order of 25 deg.

5-6 ONE-QUADRANT CONVERTERS

Two-quadrant converters can operate with both positive and negative mean dc load voltages, and will supply ac power from the source to the dc load, the rectifying mode, and will remove dc power from the load and return it to the ac source in the synchronous inversion mode.

Many applications require only a power flow from the ac source to the dc load and thus need only be operated in the rectifying mode. This is achieved in bridge converters by replacing half the thyristors in each leg by diodes, with a consequent reduction in device costs and a simplification in the firing control

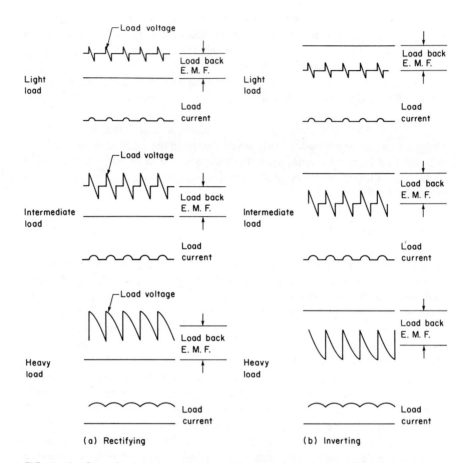

(a) Rectifying   (b) Inverting

FIG. 5-10 Six-pulse bridge converter with an active load. (a) Rectifying mode; (b) inversion mode.

circuitry. These converters are called half-controlled bridge converters.

An alternative method of obtaining one-quadrant operation in both bridge and midpoint converters is to connect a "freewheel" diode across the load terminals of the converter. Major benefits of one-quadrant operation are a reduction in the ripple content of the mean dc voltage and a reduction in the lag of the fundamental supply current with respect to the supply voltage.

5-6-1   Two-pulse Half-controlled Bridge Converter

The basic circuit of the two-pulse half-controlled bridge converter shown in Fig. 5-11(a) is in fact two separate two-pulse midpoint converters in series. The

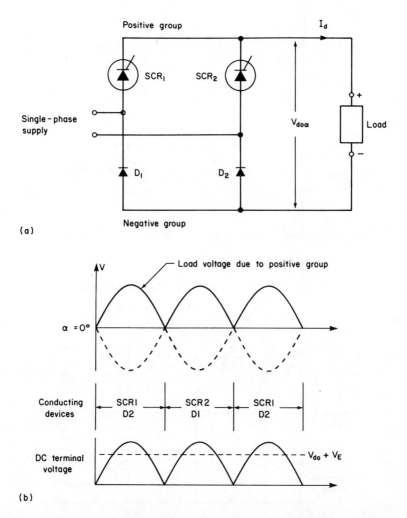

(a)

(b)

FIG. 5-11 Two-pulse half-controlled bridge converter. (a) Basic circuit (b) load voltages due to positive and negative groups and dc terminal voltage for $\alpha = 0°$.

positive group with a common cathode connection is a two-pulse midpoint converter whose output voltage is controlled by phase-shift control of SCRs 1 and 2. The second midpoint converter forming the negative group consists of diodes D1 and D2 with a common anode connection and is uncontrolled.

When the firing delay angle of the positive group is varied, the mean output dc voltage of the bridge can be varied from its maximum positive value to nearly zero as the firing delay angle is varied from 0 deg to nearly 180 deg.

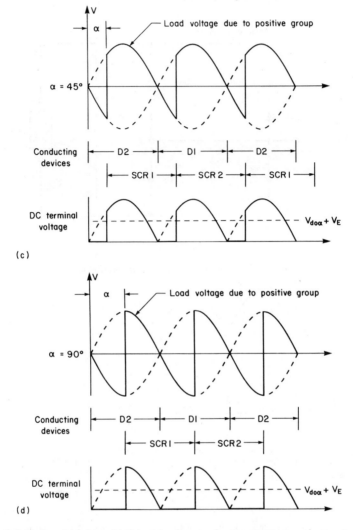

FIG. 5-11 (*cont.*) (c) and (d) load voltages due to positive and negative groups and dc terminal voltage for $\alpha = 45°$, and 90°.

When $\alpha = 0°$, SCR1 and D2 are conducting, and the voltages at the common anode and cathode connections are equal and opposite; the voltage applied to the load is twice that of either converter, with the bridge converter acting as a diode bridge [Fig. 5-11(b)].

When $\alpha = 45°$ [Fig. 5-11(c)], for the first 45 deg the voltages at the common anode and cathode connections are equal and have the same polarity, and the voltage applied to the load is zero. At 45 deg SCR1 is turned on, and the supply voltage is applied to the load. The mean dc load voltage is then the sum of the voltages across both converters, and the load current flows through SCR1 and D2. When SCR1 becomes reverse-biased, if an inductive load is assumed, SCR1 remains in conduction until SCR2 is turned on, and load current will freewheel through the path SCR1 and D1, SCR1 remaining in conduction for 180 deg. When SCR2 is turned on, SCR1 is commutated off, and the load current flows through SCR2 and D1 until SCR2 and D1 are reverse-biased. At this point SCR2 will remain in conduction, and the load current will freewheel through SCR2 and D2. During the period that the load current is freewheeling, there will be no current supplied from the ac source.

When $\alpha = 90°$, exactly the same conditions prevail [see Fig. 5-11(d)].

The firing delay angle may be increased to approximately 160–170 deg without fear of a commutation failure, but since the load voltage is the sum of the voltages across both midpoint converters, it is impossible to reduce the load voltage to zero. A major failing of this configuration occurs at small firing delay angles with a highly inductive load. Let us assume that $\alpha = 20°$ and SCR1 has turned on, and it is now decided to turn the bridge off by removing the gate signals. With a sufficiently inductive load, when SCR1 becomes reverse-biased, the load current will flow through SCR1 and D1, if the load current has not decayed to zero during the period of time SCR1 is reverse-biased. Then as SCR1 becomes forward-biased again, it will conduct as if $\alpha = 0°$, and load current will flow through SCR1 and D2 for a complete half-cycle; the converter is said to be "half-waving". The only way to turn off the bridge under these conditions is to reapply the gate signals and increase the firing delay angle, thus reducing the mean dc output voltage and load current, at which point the load current will be insufficient to maintain the SCR conducting during its reverse-biased period.

An alternative circuit arrangement shown in Fig. 5-12 eliminates this problem. The freewheeling current is restricted to the path including the two diodes in series, resulting in the periods of conduction for the diodes increasing and that of the thyristors decreasing, and in fact the thyristors block for a full 180 deg. The mean dc load voltage can be reduced to zero, and sudden removal of the gate signals does not result in "half-waving".

In summary, the negative excursions of voltage at the dc terminals experienced in the full-controlled converter at large firing angles are replaced by zero voltages as the load current freewheels, thus reducing the ripple content of the load voltage and reducing the filtering requirement.

In the case of a full-controlled bridge with an inductive load the supply current from the ac source will flow for 180 deg and will progressively lag the supply voltage as the firing delay angle is increased. In the half-controlled converter, the duration of the ac supply current decreases as the firing angle is increased, so that when the mean dc voltage is zero the supply current is zero, and as a result the apparent power factor of the half-controlled converter is improved.

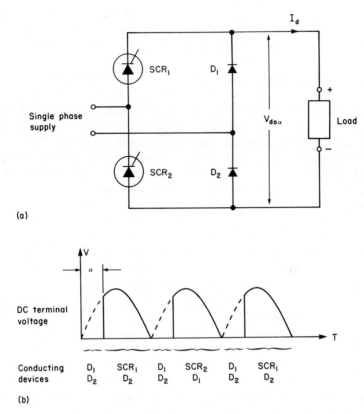

(a)

(b)

FIG. 5-12 Two-pulse half-controlled bridge converter eliminating half-waving. (a) Basic circuit; (b) dc terminal voltage.

Other possible configurations of the two-pulse bridge converter are shown in Fig. 5-13.

It should also be noted that all of the configurations shown in Figs. 5-11, 5-12, and 5-13 can operate only in the rectifying mode.

5-6-2 Three-pulse Half-controlled Converter

The basic circuit of a three-pulse half-controlled converter is shown in Fig. 5-14(a). It consists of a positive controlled three-pulse midpoint converter and a negative uncontrolled three-pulse rectifier. Variation of the firing delay angle of the controlled converter permits it to operate from rectification to full inversion, and since its voltage is additive to that of the uncontrolled rectifier, the mean dc load voltage can be varied from a positive maximum to approximately zero.

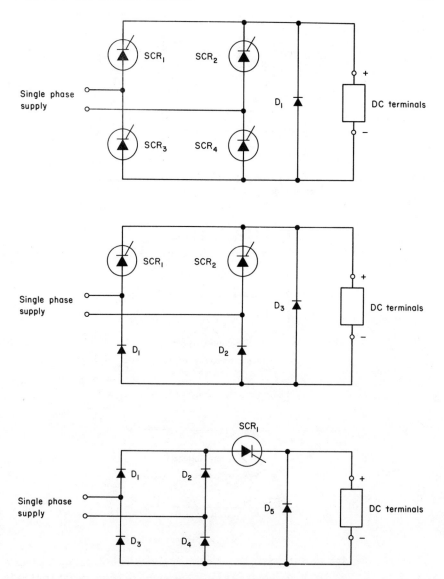

FIG. 5-13 Two-pulse bridge converters, alternative arrangements.

When $\alpha = 0°$, the bridge acts as a three-phase rectifier bridge, and the mean dc voltage V_{do} is at its maximum value with a ripple frequency of $6f$. Each device conducts for 120 deg, and the ac supply current consists of alternating components of 120 deg in phase with the supply voltage [see Fig. 5-14(b)].

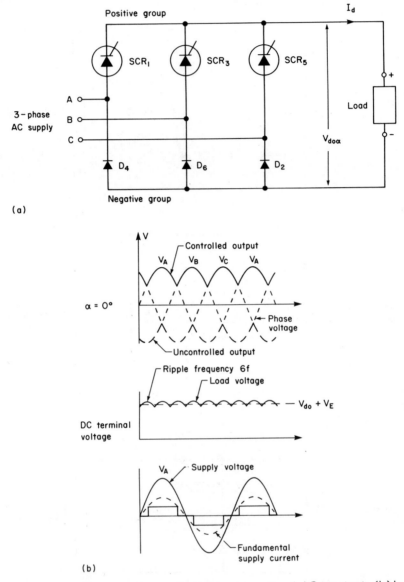

FIG. 5-14 Three-pulse half-controlled bridge converter. (a) Basic circuit; (b) load voltages due to controlled and uncontrolled groups, dc terminal voltage for $\alpha = 0°$.

When $\alpha = 60°$, the mean dc voltage $V_{do\alpha}$ is the difference voltage between the positive and negative converters. However, the ac input current to the positive converter has been retarded 60 deg and then flows for 120 deg, while the

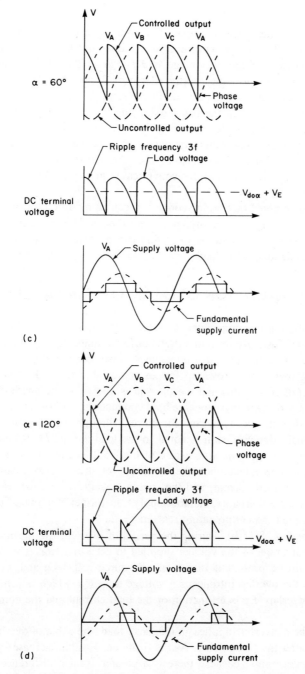

FIG. 5-14 (*cont.*) (c) and (d) load voltages due to controlled and uncontrolled groups, dc terminal voltage for $\alpha = 60°$, and $120°$.

ac current due to the negative diode converter is unaltered in duration and position. The resulting supply current is still alternating. It should be noted that the ripple frequency of the dc voltage is $3f$ at $\alpha \geqslant 60°$. This is due to the appearance of a third harmonic shortly after $\alpha = 0°$ and, becomes pronounced after $\alpha = 30°$. The thyristor and diode in the same leg are both in conduction when $\alpha = 60°$; increases in the firing delay angle will result in this condition continuing with the current freewheeling through the thyristor and diode, thus reducing the mean dc voltage and supply current to zero during the freewheeling periods [see Fig. 5-14(c)]. It can also be seen that the filtering requirements of the three-pulse half-controlled bridge converter are greater than those of the full-controlled converter.

It should be noted that bridge converters can be modified for one-quadrant operation by replacing half the thyristors with diodes. However, in the case of midpoint converters one-quadrant operation can be obtained only by shunting a freewheel diode across the load.

5-7   COMMUTATION OVERLAP, OVERLAP ANGLE μ

In all the configurations considered to date, it has been assumed that the load I_d is commutated instantly from one thyristor to the other. In actual practice this is not the case, since the ac source impedance causes conduction in the outgoing and incoming thyristors to overlap for an appreciable time. This situation is called *commutation overlap*, and the interval is known as the angle of overlap μ. During this period the thyristors are both conducting, with the current through the outgoing thyristor decaying and the current through the incoming thyristor increasing. Their algebraic sum is equal to the load current I_d, which remains constant because of the inductive nature of the load. The duration of the angle of overlap μ is dependent upon the amplitude of the ac source voltage, the magnitude of the load current, and the source reactance, i.e., the reactance of the supply lines, the thyristors, and the supply transformer windings. This source reactance is called the commutating reactance, X_c.

Consider Fig. 5-15(b). During the period of overlap, the mean dc output voltage is the average of the voltage supplied to each thyristor. This voltage is less than would be present if there was no source reactance and with a zero overlap, i.e., the overlap introduces a voltage drop E_x, which is a function of the angle of overlap. E_x is a function of the load current and the commutating reactance.

When the converter supplies an inductive load, the angle of overlap occurs immediately after the firing delay. As the firing delay angle increases, the angle of overlap μ decreases, because there is a greater voltage difference between the phases, which increases the rate of commutation of the load current from one thyristor to the other.

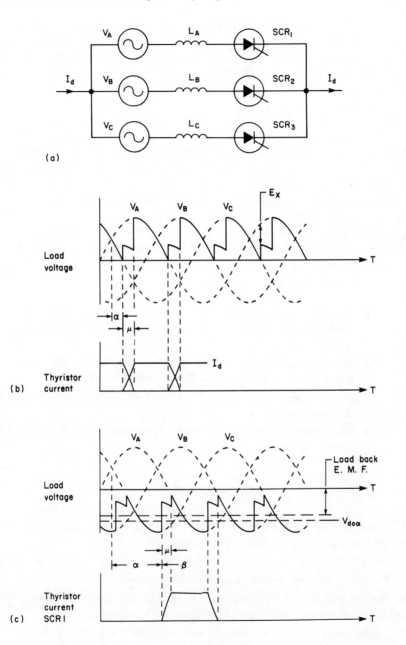

FIG. 5-15 Commutation overlap, three-pulse midpoint converter with finite commutating reactance. (a) Basic circuit; (b) voltage and current waveforms, rectifying; (c) voltage and current waveforms, inverting.

When the converter is operating in the inversion mode [Fig. 5–15(c)], that is, the firing delay angle α is greater than 90 deg, the mean dc output voltage $V_{do\alpha}$ is negative and is less than the dc source voltage V by the amount necessary to drive the load current I_d through the load. Commutation occurs when the incoming thyristor is triggered and its anode is more positive than the outgoing thyristor anode. The voltage difference between the phases produces a commutating current that reduces the current through the outgoing thyristor and promotes current buildup in the incoming thyristor. If for any reason the first delay angle is greater than 180 deg, the commutating voltage is no longer present and conduction is now occurring in the positive half-cycle. This voltage aids the dc source voltage, and a short circuit will occur. For correct operation, commutation must take place before the phase voltages become equal. The angle by which the firing point must be advanced is $\beta = 180° - \alpha - \mu$ in order that sufficient time is allowed for overlap and turn-off. It should be noted that overlap results in an increase in the back emf of the converter in the inversion mode, with a resultant increase in the load current I_d.

In summary it should be noted that:

1. The mean dc voltage $V_{do\alpha}$ is decreased by overlap in the rectifying mode, and the firing delay angle must be advanced to maintain the same output voltage.
2. The ac source voltage will be distorted, which may affect the firing signal synchronization of the firing control circuit.
3. The inverter advance angle β will be reduced when operating in the inversion mode.
4. The harmonic content of the mean dc voltage will be increased when overlap is present.

REVIEW QUESTIONS

1. What is meant by synchronous inversion?
2. What is the significance of the pulse number?
3. What is meant by the following terms: a one-quadrant converter; a two-quadrant converter; a dual or four-quadrant converter?
4. Discuss with the aid of a schematic and waveforms the principle of operation of a single-phase, half-wave phase-controlled converter.
5. Explain with the aid of a schematic and waveforms the principle of operation of a two-quadrant, two-pulse midpoint converter in the rectifying and inverting modes of operation.
6. Explain with the aid of a schematic and waveforms the principle of operation

of a two-quadrant, two-pulse bridge converter in the rectifying and inverting modes of operation.

7. What are the advantages and disadvantages of midpoint converters versus bridge converters?

8. Explain the principle of operation of a three-pulse midpoint converter.

9. What are the advantages and limitations of using a six-pulse midpoint converter with an interphase reactor, as compared to a six-pulse midpoint converter?

10. Discuss with the aid of a schematic and waveforms the principle of operation of a six-pulse bridge converter.

11. What is meant by discontinuous conduction? Illustrate your answer with load current waveforms for passive R-L and active loads.

12. What are the advantages and disadvantages of using one-quadrant converters?

13. What is meant by "half-waving", and how may it be eliminated?

14. What is the function of a freewheel diode?

15. With the aid of a schematic and waveforms discuss the operation of a three-pulse half-controlled converter.

16. What is meant by commutation overlap? What is the cause of commutation overlap? What is the effect of commutation overlap on the performance of a converter operating in the rectifying and inversion modes?

6 Static Frequency Conversion

6-1 INTRODUCTION

The requirement for an adjustable speed drive is that it provide continuous and accurate speed control with high stability and good transient performance. Traditionally this requirement has been met by closed-loop adjustable speed dc drives using armature voltage control techniques based on the Ward-Leonard system. These drives have held a predominant position in machine tool control, steel and paper mills, variable delivery pumps and fans, mine hoists, and cranes.

The introduction of gaseous devices such as the ignitron and mercury pool rectifiers in the 1940s permitted the application of electronics to the speed control of dc machines by means of variable armature voltage.

However, in spite of the control capabilities of dc drives, the initial costs of a system, combined with the ongoing maintenance costs, have placed the dc drive at an economic disadvantage as compared to the variable frequency control of induction and synchronous motors.

Although the principle of variable frequency static inverters and cyclo-converters has been known for a long time, it has been economically feasible to develop these drives only with the introduction of the thyristor. Decreasing thyristor costs and the development of low-cost digital and analog integrated circuits have provided a tremendous boost to the development of variable frequency ac drives. This development, combined with an initially lower motor cost and substantially reduced maintenance costs, is continually making the ac drive the logical choice to replace the static dc drive.

Other factors that weight the selection of a drive in favor of ac are;

1. Many drives can accept frequencies up to 200 Hz, resulting in a potential 12,000 rpm with a two-pole machine.

2. Speed control ranges from 6:1 up to approximately 20:1 are obtainable.

3. The squirrel cage rotor has a relatively low inertia, which permits a better dynamic response.

4. A wide range of standard motors are available from stock.

5. Speed regulation is improved, since the output frequency is not load dependent.

6. Open-loop control provides excellent speed control without the added costs of speed-measuring equipment and error-correcting regulators required by closed-loop control methods.

7. Multimotor drive systems are easily synchronized.

8. Four-quadrant operation is readily achieved.

6-2   BASIC POLYPHASE INDUCTION MOTOR THEORY

Before we discuss static frequency conversion methods, it is necessary to establish the basic requirements of such systems.

When a balanced three-phase supply is applied to the stator of a three-phase induction motor, a constant amplitude rotating magnetic field is produced. The angular velocity of this field is given by

$$S = 120\,\frac{f}{P}\ \text{rpm} \qquad\qquad (6\text{-}1\text{E})$$

or
$$\omega = 4\pi\,\frac{f}{P}\ \text{rads/sec} \qquad\qquad (6\text{-}1\text{SI})$$

where S or ω = synchronous speed of the rotating magnetic field in the appropriate units
   f = frequency, Hz
   P = number of stator poles/phase

from which it can be seen that the synchronous speed of the rotating field for a given machine is

$$S\,\alpha\,f \qquad\qquad (6\text{-}2)$$

The rotor of a normal induction motor runs at a speed slightly less than the synchronous speed of the rotating magnetic field. The difference from the synchronous speed is called the *slip* and is usually expressed as a percentage. Slip is defined as

$$s = \frac{\text{synchronous speed-rotor speed}}{\text{synchronous speed}} \quad\quad (6\text{-}3E)$$

or
$$s = \frac{(S - S_r)100}{S} \text{ percent}$$

or in SI units
$$s = \frac{(\omega - \omega_r)100}{\omega} \text{ percent} \quad\quad (6\text{-}3SI)$$

where s = percent slip or decimal slip without 100
S or ω = synchronous speed
S_r or ω_r = actual rotor speed

Under normal operating conditions the ac supply voltage and frequency are constant. When the rotor is stationary, the synchronous rotating magnetic field will induce emfs at supply frequency in it. If the short-circuited rotor conductors are purely resistive, the rotor emf and rotor current will be in phase, and the torque contribution of each rotor conductor is proportional to the product of the flux density at the conductor and the conductor current; that is, the conductor with the greatest current will be exposed to the greatest air-gap flux density and maximum electromagnetic torque will result. If the rotor circuit is purely inductive, the resulting current will lag the induced emf by 90 deg, and the electromagnetic torque will be zero. Normally, the rotor power factor is greater than zero, and an electromagnetic torque will be present, which will accelerate the rotor from standstill in the same direction as the rotating magnetic field, to a speed less than the synchronous speed of the field.

The frequency of the induced rotor voltage is

$$f_r = \left(\frac{S - S_r}{S}\right)f = sf \quad\quad (6\text{-}4E)$$

or
$$f_r = \left(\frac{\omega - \omega_r}{\omega}\right)f = sf \qu\quad (6\text{-}4SI)$$

Under normal running conditions the rotor frequency is normally low, 2 to 5 Hz, and the rotor reactance is small. The rotor current is limited by the rotor resistance. The air-gap flux can also be considered as constant. The induced rotor emf will be proportional to the fractional slip. If the rotor rotates in synchronism with the rotating magnetic field, i.e., $s = 0$, the induced rotor emf is zero and no torque is produced. As a result, even at no-load the rotor speed is less than the synchronous speed of the rotating magnetic field, since torque must be provided to overcome the windage and friction losses of the motor. As the mechanical loading on the motor is increased, the rotor speed drops and the slip increases, with a resulting increase in the induced rotor emf, rotor current, and electromagnetic torque, until an equilibrium is reached.

6-2-1   Equivalent Circuit of a Polyphase Induction Motor

The performance of a polyphase induction motor can be predicted under steady-state conditions from an analysis of the equivalent circuit shown in Fig. 6-1.

The polyphase induction motor is basically a polyphase transformer with a rotating secondary winding. The stator current produces a mutual flux, which links with the rotor winding, and a stator leakage flux, which links with the stator winding only. The leakage flux induces a stator counter emf. The applied stator emf, V_1, is greater than the counter emf, E_1, by the voltage drop in the stator leakage impedance, which on a per-phase basis is

$$V_1 = E_1 + I_1(r_1 + jx_1) \tag{6-5}$$

where V_1 = applied stator voltage
E_1 = counter emf produced by the resultant air-gap flux
I_1 = stator current
r_1 = effective stator resistance
x_1 = stator leakage reactance

The air-gap flux is the resultant of the fluxes produced by the stator and rotor currents. The stator current can be resolved into two components, the load current I_2, which produces an emf that exactly counterbalances the emf produced

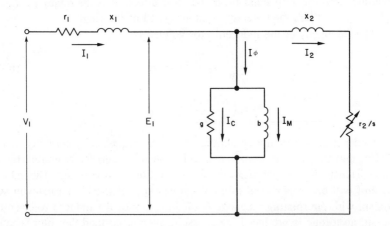

FIG. 6-1   Equivalent circuit of a polyphase induction motor on a per-phase basis.

by the rotor current, and a no-load current I_ϕ consisting of the magnetizing component I_m and the core loss component I_c. The rotor leakage reactance is dependent upon the frequency of the rotor currents, i.e., $f_r = sf$; therefore, the rotor leakage reactance is sx_2, where x_2 is the locked rotor leakage reactance. The rotor emf at standstill in the equivalent circuit is equal to the stator emf; therefore, at slip s, the rotor emf $E_2 = sE_1$. If in the equivalent circuit r_2 is the rotor resistance per phase and x_2 is the rotor standstill reactance per phase then the rotor current I_2 is

$$I_2 = \frac{E_2}{r_2 + jsx_2} = \frac{sE_1}{r_2 + jsx_2} \tag{6-6}$$

in terms of slip frequency, thus

$$I_2 = \frac{E_1}{(r_2/s) + jx_2} \tag{6-7}$$

in terms of the supply frequency.

6-2-2   Power Balance Equations

From the equivalent circuit of Fig. 6-1 for an m phase machine,

$$\text{Input power} = mV_1I_1 \cos \phi_1$$

$$\text{Stator copper loss} = mI_1{}^2r_1 \tag{6-8}$$

$$\text{Rotor power input, RPI} = mI_2{}^2r_2/s \tag{6-9}$$

$$\text{Rotor copper loss, RCL} = mI_2{}^2r_2 \tag{6-10}$$

$$\text{Rotor power developed, RPD} = (6\text{-}9) - (6\text{-}10) = \text{RPI} - \text{RCL}$$

$$= mI_2{}^2r_2\left(\frac{1-s}{s}\right) \tag{6-11}$$

and $\text{RPD} = \text{RPO} + P_{\text{ROT}}$

where   RPO = rotor mechanical power output
         P_{ROT} = rotational losses, i.e., windage and friction

If ω_r is the angular velocity of the rotor and T is the electromagnetic torque, then

$$Tw_r = RPD = mI_2^2 r_2\left(\frac{1-s}{s}\right)$$

(6-12)

and
$$T = \frac{mI_2^2 r_2}{w_r}\left(\frac{1-s}{s}\right)$$

where T is the internal motor torque and is greater than the actual shaft torque by the amount required to overcome the rotational losses.

Since the angular velocity of the rotating magnetic field is

$$\omega = \frac{w_r}{1-s} = \frac{4\pi f}{P}$$

then Eq. (6-12) can be rewritten as

$$T = \frac{mI_2^2 r_2}{w_r}\left(\frac{1-s}{s}\right) = \frac{mI_2^2 r_2}{\omega s}$$

$$= \frac{mP}{4\pi f} \cdot I_2^2 \cdot \frac{r_2}{s}$$

(6-13)

From
$$I_2 = \frac{E_1}{r_2/s + jx_2} = \frac{E_1}{\sqrt{[(r_2/s)^2 + x_2^2]}}$$

and
$$s = f_r/f$$

then
$$T = \frac{mP}{4\pi} \cdot \left[\frac{E_1}{f}\right]^2 \cdot \frac{f_r r_2}{[r_2^2 + s^2 x_2^2]}$$

(6-14)

The air-gap flux must be maintained constant at all frequencies to produce a constant torque. When the ratio E_1/f is constant, a constant air-gap flux is produced. If the stator leakage impedance is small, then $E_1 \simeq V_1$, and the air-gap flux is approximately constant when the V_1/f ratio is constant. This is known as the constant volts/Hz method of operation and is most commonly used in open-loop control of static inverters and cycloconverters.

However, at low frequencies, since the stator resistance is the predominant component (see Fig. 6-2), there will be a decrease in air-gap flux at low frequencies with a consequent reduction in motor torque. If the low-speed performance is not acceptable with a constant voltage/Hz ratio, then the control must be modified to increase the volts/Hz ratio at low frequencies. This action will not cause an undue increase in core losses, even though the stator iron may be saturated.

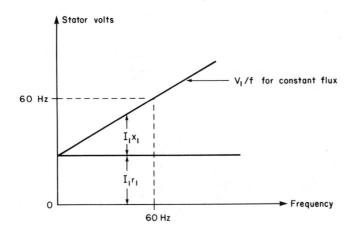

FIG. 6-2 Stator voltage versus frequency.

Equation (6-14) may be written in terms of the rotor frequency as

$$T = \frac{mP}{4\pi}\left[\frac{E_1}{f}\right]^2\left[\frac{f_r r_2}{r_2^2 + (2\pi f_r l_2)^2}\right] \tag{6-15}$$

and differentiating with respect to rotor frequency f_r and equating to zero gives the relation for the rotor breakdown frequency as

$$f_b = \pm \frac{r_2}{2\pi l_2} \tag{6-16}$$

where f_b = the rotor frequency at which maximum torque occurs. Substituting in Eq. (6-15), we find that the maximum or breakdown torque is

$$T_b = \pm \frac{mP}{4\pi}\left[\frac{E_1}{f}\right]^2 \cdot \frac{1}{4\pi l_2} \tag{6-17}$$

where f_b = rotor breakdown frequency
         T_b = torque at breakdown
         l_2 = rotor leakage inductance
         − sign implies the motor is operating as a generator

Since the air-gap flux is proportional to E_1/f, then the breakdown torque is proportional to air-gap flux squared, and inversely proportional to the rotor leakage inductance. It should be noted that the rotor resistance does not affect the magnitude of the breakdown torque, but from Eq. (6-16), it will affect the rotor frequency at which breakdown occurs.

It is very often convenient to express torque in terms of the torque ratio t, which is the ratio of the torque T at any rotor frequency f_r to the maximum torque T_b at the rotor breakdown frequency f_b. Substituting Eqs. (6-16) and (6-17) in Eq. (6-15) produces

$$t = \frac{T}{T_b} = \frac{2}{(f_r/f_b) + (f_b/f_r)} \tag{6-18}$$

for constant flux operation. Also for a given machine

$$T_b = \text{a constant} \tag{6-19}$$

When a polyphase induction motor is operated under constant flux conditions, the torque supplied is greater at all supply frequencies than when supplied at rated voltage and frequency.

As a result, a variable frequency induction motor drive will have a greater starting torque at low frequencies, and for the same operating slip, the torque will be greater at higher frequencies. For a given load torque, the horsepower output is proportional to frequency, and as the frequency is increased, the efficiency increases, since

$$\eta = \frac{\text{RPO} - P_{\text{ROT}}}{\text{RPI} + \text{stator copper losses} + \text{core losses}} \tag{6-20}$$

$$\simeq \frac{\text{RPO}}{\text{RPI}} \simeq \frac{m I_2^2 r_2 \cdot \dfrac{1-s}{s}}{m I_2^2 r_2 / s}$$

$$\simeq (1 - s) \tag{6-21}$$

From Fig. 6-3, at a constant load torque it can be seen that as the supply frequency is increased, the slip s decreases, and thus the motor efficiency increases.

In summary, the operation of a polyphase induction motor, either squirrel cage or wound rotor, with a constant volts/Hz ratio, results in a higher starting and breakdown torque, and with the same full load slip the torque is greater at higher frequencies than at the lower frequencies. In addition, the horsepower (kW) output and efficiency are greater at the higher frequencies. Failure to maintain the constant volts/Hz ratio will either affect the constant torque by the square of the air-gap flux density, or permit the stator current to increase and overheat the motor.

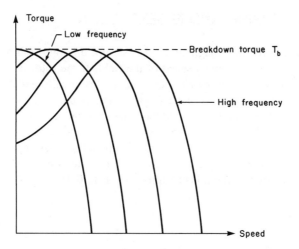

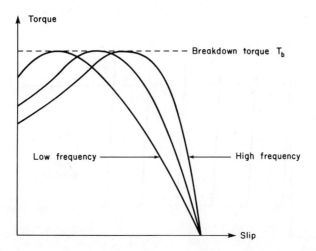

FIG. 6-3 Torque-speed and torque-slip curves of a polyphase induction motor with a variable-frequency source.

Because a solid-state static frequency converter can be programmed to provide the optimum voltage and frequency to a polyphase induction motor for any desired operating speed, there is no requirement for reduced voltage starting equipment; therefore, starting inrush currents are eliminated, and the maximum torque per kVA is developed.

6-3   MOTORING AND REGENERATION

When a polyphase rotating magnetic field is set up, the rotor will turn at some speed less than the synchronous speed of the field. If the rotor is connected to an overhauling load, the rotor will accelerate and eventually exceed the speed of the synchronous rotating magnetic field. At this point, the rotor conductors are sweeping past the rotating field and the induced rotor emfs and rotor currents are reversed, the reflected stator currents are also reversed, and the motor acts as an induction generator. When the motor is operating as an induction generator, regeneration is occurring and the overhauling load is being slowed by the energy being fed back to the power supply (see Fig. 6-4).

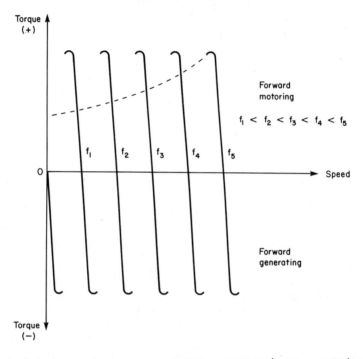

FIG. 6-4 Polyphase induction motor, torque versus speed curves, motoring, and regeneration with a variable-frequency supply.

6-4   FORWARD AND REVERSE ROTATION

In order to reverse a polyphase induction motor, it is necessary to reverse any pair of the stator leads. In addition, provided that the rotor can be accelerated through synchronous speed the motor will operate as an induction generator in the reverse direction also; that is, the polyphase induction motor can be operated in all four quadrants with only the necessity of changing the phase sequence of the supply to the stator. The only other drive that has this capability is a dc Ward-Leonard system supplied by a two-quadrant converter with a reversing switch at its output or a dual converter (four-quadrant).

6-5 STATIC FREQUENCY CONVERSION VERSUS A VARIABLE FREQUENCY MOTOR-ALTERNATOR DRIVE

Prior to the use of static frequency converters, variable frequency, variable voltage was provided by variable-speed motor-alternator sets. This system had two major limitations:

1. The motor-alternator speed has to be changed to produce a frequency change with a resulting poor dynamic response for the combined motor-alternator, induction motor drive.

2. The amplitude of the alternator voltage is proportional to the rate of cutting magnetic lines of force, which means that at low speeds, in spite of raising the alternator field excitation to a maximum, the amplitude of the output would be very small, and the unit would be incapable of providing the constant voltage/Hz required for constant torque operation.

It can be seen from the above that the initial and ongoing maintenance costs combined with the performance limitations of the variable-speed motor-alternator system very quickly eliminates it as a competitor to a static frequency conversion system with a constant torque capability, fast response, and open-loop control techniques.

6-6 THE DC-LINK CONVERTER

The dc-link converter, which is one of the two major methods of obtaining a variable-frequency supply, consists of a dc rectifier and an inverter. Usually, the dc rectification is accomplished by a three-phase, phase-controlled thyristor bridge converter, or a three-phase diode bridge whose dc output is supplied to a three-phase inverter. Voltage control to obtain a constant volts/Hz relationship can be obtained by phase angle control of the phase-controlled converter or by voltage control internally in the inverter.

6-6-1   Single-phase Bridge Inverter

The principle of inverter operation is best introduced by considering the single-phase inverter shown in Fig. 6-5. It can be seen immediately that only one thyristor in each leg can be on at any one time; i.e., SCR1 or SCR2 may be on,

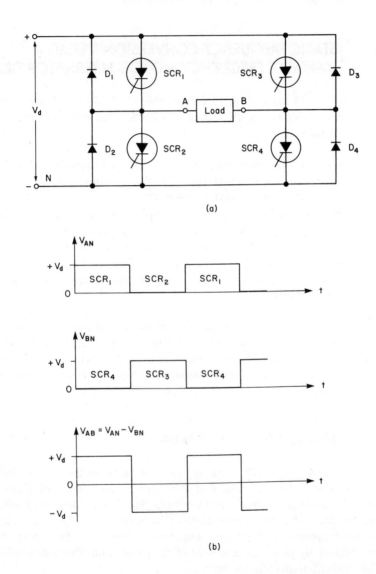

(a)

(b)

FIG. 6-5   Single-phase bridge inverter. (a) Basic circuit; (b) voltage waveforms.

but not both simultaneously, or a short circuit will be applied across the dc source. The thyristors are connected in a bridge configuration with the load connected between points A and B. If the load is reactive, i.e., either inductive or capacitive, a return path for the reactive energy must be provided to the dc source. The function of the feedback diodes connected in inverse-parallel with the thyristors is to provide a path for the reactive energy.

The operation of the inverter is as follows: All SCRs are gated on and turned off in the appropriate sequence for a period of time corresponding to 180 deg of the output ac cycle. Then if SCR1 is turned on, point A will be positive with respect to the negative bus. When SCR1 is turned off and SCR2 is turned on, point A will be at the potential of the negative bus. Similarly, when SCR3 is on, point B is at the positive bus potential. Turning off SCR3 and gating on SCR4 places point B at the negative bus potential. Thus cycling SCRs 1 and 2 on and off alternately produces a series of positive pulses of voltage V_{AN}; cycling SCRs 3 and 4 produces a series of positive pulses of voltage V_{BN}, displaced 180 deg from those of V_{AN}.

If now SCRs 1 and 4 are gated on for 180 deg and then turned off, and SCRs 2 and 3 are gated on for 180 deg and then turned off, the voltage applied to the load, V_{AB}, is produced, which is an alternating square wave of peak amplitude V_d.

In the case of a pure resistive load, the load current I_L will have exactly the same waveform and be in phase with the load voltage. If the load is inductive, the load current will lag the inverter output voltage. The function of the feedback diodes is to return the reactive energy from the load back to the dc source; for example, if SCRs 1 and 4 are turned off, the reactive energy will return via diodes D2 and D3; when SCRs 2 and 3 are turned off, the reactive energy is returned via diodes D1 and D4.

Another function of the diodes is to prevent the peak amplitude of the inverter output exceeding that of the dc source, and as a result the output voltage will always have a constant amplitude.

The gate firing circuits and the commutation circuitry for the SCRs have been deliberately omitted at this point to simplify the circuit explanation.

6-7   SINGLE-PHASE INVERTER VOLTAGE CONTROL

The basic concepts of inverter output voltage control can be most easily understood by considering the single-phase inverter. Most inverter applications require a means of controlling the output voltage. This may be achieved by either controlling the dc voltage supplied to the inverter, or by controlling the voltage within the inverter.

6-7-1   Variable-input Voltage Control

The application of dc link converters to variable-frequency ac motor drive systems requires a constant volts/Hz ratio in order that a constant torque output can be obtained.

One method of controlling the volts/Hz ratio is by variation of the dc input voltage to the inverter input terminals.

If the source voltage is ac, then the dc input voltage to the inverter may be controlled by a number of methods, namely:

1. A phase-controlled converter.
2. An uncontrolled rectifier with a variable dc output voltage being obtained by:
   a. Variation of the input ac voltage by means of an induction regulator or a variable autotransformer. Either of these methods requires mechanical adjustment of the device and results in a poor dynamic response for the overall inverter system, and increased control complexity.
   b. Variation of the output dc voltage by means of a dc-dc controller or chopper.

If the source voltage is dc, then the major method of varying the dc input voltage to the inverter is by means of a chopper.

As can be seen, there are a number of methods of obtaining a variable dc voltage at the inverter input terminals; however, the system in most common use is the thyristor phase-controlled converter.

The major advantage of controlling the dc input voltage to the inverter input terminals is that the inverter output waveform and its harmonic content do not vary greatly as the dc voltage level is changed. However, the disadvantages outweigh the advantages, namely:

1. The current-commutating capability of the inverter varies with the dc voltage level. This is because a reduction in voltage without a corresponding reduction in the current reduces the time available for turn-off. This problem can be offset to some extent by increasing the commutating capacitance or by providing a fixed dc commutating supply with a resulting increase in circuit complexity.
2. The control of the dc voltage adds to the complexity of the overall dc link converter.
3. The dc voltage should be smooth, which requires filtering of the dc output voltage, which in turn reduces the response time of the system.

6-7-2   Voltage Control within the Inverter

Because of the disadvantages of the variable voltage input control techniques, the most efficient method of voltage control is by varying the ratio between the dc voltage at the inverter input terminals and the ac output voltage of the inverter. The methods in common usage, in general, differ only because of the effect that they have on the harmonic content of the output ac voltage waveform, with the harmonic content increasing as the output voltage is decreased. The most common methods of internal voltage control all use pulse width modulation techniques and are:

1. Pulse width control.
2. Pulse width modulation (PWM).

6-7-2-1   Pulse Width Control

In this technique the inverter output voltage is varied by controlling the width of the output pulse. In the circuit of Fig. 6-5, if SCRs 1 and 4 are turned on for one half-cycle, and SCRs 2 and 3 are turned on for the next half-cycle, the output voltage waveform shown in Fig. 6-6(a) will result. Voltage control can be obtained by varying the phase relationship between the firing of SCRs 3 and 4 with respect to SCRs 1 and 2. The effect of retarding SCRs 3 and 4 90 deg with respect to SCRs 1 and 2 is shown in Fig. 6-6(b). It can also be seen that varying the retardation angle γ from 0 to 180 deg will vary the inverter output voltage from a maximum to zero, with only one commutation of each SCR for each cycle of the inverter output voltage, thus reducing the commutation losses to a minimum.

The major disadvantage of pulse width control is that as the output pulse width δ decreases, the mean output voltage decreases and the harmonic content of the output voltage waveform increases, until at the point where the fundamental is 20 percent of its maximum, the third, fifth, and seventh harmonics are nearly equal in value to the fundamental. The problem of the high harmonic content at reduced output levels can be greatly reduced by using pulse width modulation.

6-7-2-2   Pulse Width Modulation (PWM)

Instead of reducing the pulse width, as is done in pulse width control to obtain voltage variations, the output of the inverter is switched on and off rapidly a number of times during each half-cycle, producing a train of constant amplitude pulses.

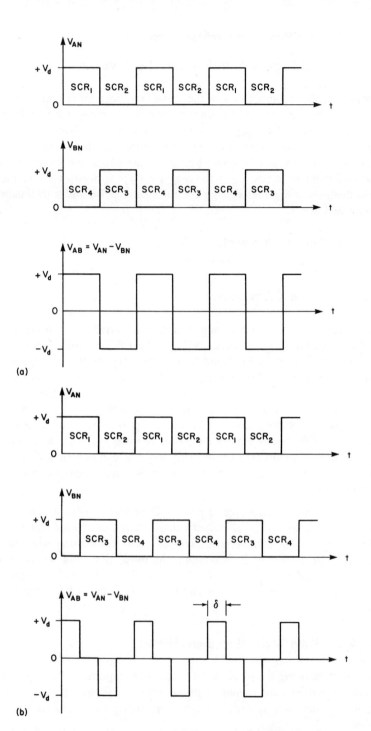

FIG. 6-6 Pulse width voltage control. (a) 0° retard; (b) 90° retard.

The amplitude of the inverter output voltage is controlled by varying the ratio of the total on-time of the pulses to total off-time. There are two basic approaches: first, by maintaining a constant pulse width and varying the number of pulses per half-cycle or by varying the pulse width for a constant number of pulses per half-cycle. The repetition rate of the pulses is known as the carrier frequency. The carrier frequency can be synchronized to the fundamental inverter frequency, or it can be independent of the fundamental inverter frequency.

The production of constant width pulses is accomplished by comparing a variable dc voltage level against a synchronized triangular carrier signal [see Fig. 6-7(a)]. However, the inverter output voltage still contains a significant harmonic content.

A reduction in harmonic content is obtained by varying the pulse widths in a cyclic manner. This is accomplished by comparing a sine wave of the same frequency as the inverter output against a synchronized triangular carrier [see Fig. 6-7(b)].

The control techniques to obtain constant width pulses are simpler than

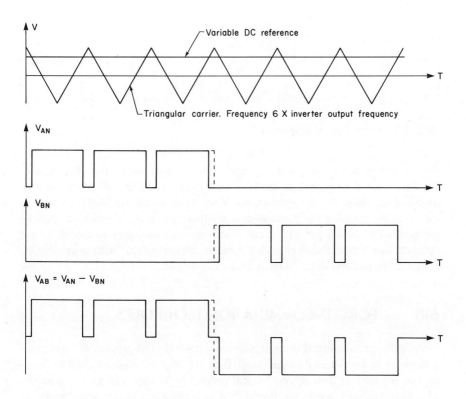

FIG. 6-7 Pulse width modulation by comparing a triangular carrier with (a) a variable dc reference.

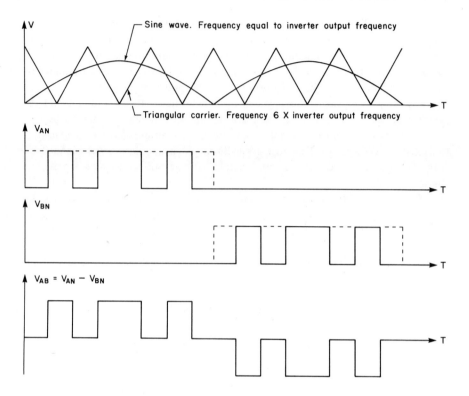

FIG. 6-7 (cont.) (b) a sine wave.

those required for varying pulse widths. The ratio of the carrier frequency to the fundamental frequency of the inverter output should be maintained as high as possible. However, the turn-on and turn-off times of the SCRs, as well as the resetting time for the commutation components, place a practical limit on the number of pulses per half-cycle. The commutation losses in the SCRs and commutating circuits also rise as the pulse rate increases, with a consequent reduction in the inverter efficiency and kVA output.

6-8   FORCED-COMMUTATION TECHNIQUES

Once a thyristor is turned on, the gate loses control. The process of turning off a thyristor is known as *commutation*. Turn-off may be accomplished by interrupting the current flow by mechanical means, or by applying a reverse anode-to-cathode voltage across the thyristor. The application of a reverse anode-to-cathode voltage reduces the forward current to zero and establishes a reverse current, which sweeps the current carriers from the J_1 and J_3 end junctions. At this point the forward current has decreased to zero, and the reverse-biased end junctions block the reverse voltage. Forward blocking capability is not re-established until the current carriers have recombined at the J_2 junction.

Circuits using thyristors are classified in terms of the method of obtaining commutation. Natural commutation occurs in ac-supplied circuits. In dc-supplied circuits, the forward current must be reduced to zero by auxiliary components. This process is called *forced commutation*, and the circuits and components achieving forced commutation are called the *commutation circuits*.

Commutation in variable-frequency dc-link converters is usually achieved by impulse techniques. Impulse commutation requires the production of an impulse that reverse-biases the anode-cathode of an SCR to achieve turn-off. The amplitude of the pulse must be great enough to decrease the forward current to zero, and of sufficient duration that the forward blocking capability is restored. The pulse is usually produced by an oscillatory inductance-capacitance (*LC*) network, whose natural period of oscillation is related to the SCR turn-off time, the dc source voltage, and the maximum value of load current to be commutated.

In our considerations of impulse commutation methods the following assumptions are made to simplify the explanations:

1. That the load is inductive, which is reasonable since the text is devoted to motor control, and that the load current remains substantially constant during the commutation period. This is valid, since typical turn-off times vary between 10 to 80 microseconds.

2. That the thyristor turn-on time is very small and can be neglected, which is reasonable since inverter-grade thyristors are being used with turn-on times typically between 4 to 8 microseconds, and that *di/dt* is limited by the circuit components.

3. That the forward voltage drop of the thyristor is negligible.

Two methods of impulse commutation will be considered as applied to inverters.

6-8-1   Auxiliary Impulse-commutated Inverter

In this method of forced commutation the commutating pulse is produced by circuitry separate from the main power source. The major benefit of this method is that a loss of the gating signals will not cause commutation failure, provided that the auxiliary commutating circuit turns off the last conducting SCR. The method that will be described is also called the McMurray circuit.

With reference to Fig. 6-8, SCRs 1 and 2 are the main load-carrying thyristors; SCRs 1A and 2A are called the auxiliary thyristors and are used to

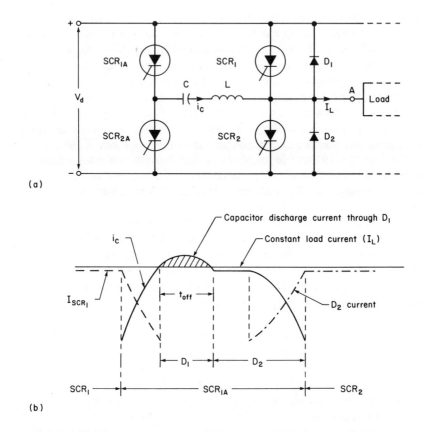

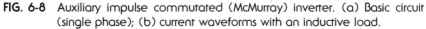

FIG. 6-8 Auxiliary impulse commutated (McMurray) inverter. (a) Basic circuit (single phase); (b) current waveforms with an inductive load.

commutate SCRs 1 and 2 by switching the high-Q LC pulse-generating network in parallel with the conducting SCR.

The operation of the circuit is as follows: It is assumed that SCR1 is in conduction and that the right-hand plate of the capacitor is positive. SCR1 is turned off by gating SCR1A on; the discharge current pulse flows through SCR1A, C, and L and opposes the load current I_L flowing through SCR1, initiating turn-off. As the discharge current i_c builds up to exceed I_L, the excess current flows through the feedback diode D1, thus applying a reverse-bias voltage across SCR1 equal to the drop across D1. The discharge current i_c peaks when the capacitor voltage has decreased to zero, and then decreases as the capacitor begins to recharge with reversed polarity. At the same time, when i_c becomes less than I_L, current flow through D1 ceases and the reverse-bias is removed from SCR1.

After SCR1 is turned off, load current is transferred to D2 if the load is inductive. SCR2 may be turned on any time after SCR1 achieves its forward blocking capability. At the same time, if it is assumed that SCR1A is still conducting, a small charging current flows through SCR1A, C, L and SCR2 to complete the charging of capacitor C with its left-hand plate positive. At this point SCR1A turns off, and SCR2 is ready to be commutated when SCR2A is fired.

The current relationships during the commutation cycle are shown in Fig. 6-8(b).

In the McMurray circuit the no-load commutating losses are small, and the capacitor voltage increases with load current, thus ensuring an improved commutating capability. This commutation method is capable of being used over a range of frequencies up to approximately 5 kHz, thus ensuring its suitability with pulse width modulation voltage control techniques.

6-8-2   Complementary Voltage Commutation, McMurray-Bedford Circuit

In the McMurray-Bedford circuit shown in Fig. 6-9(a) the turning on of SCR2 turns off SCR1, and vice versa. This technique is known as complementary

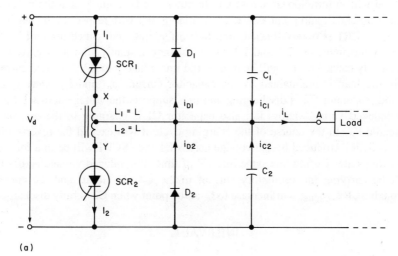

(a)

FIG. 6-9 One phase of a McMurray-Bedford commutated inverter. (a) Basic circuit.

commutation and has the advantage compared to the McMurray circuit of eliminating the auxiliary thyristors, but at the expense of additional inductance and capacitance.

In Fig. 6-9(a) the commutating capacitors C1 and C2 are equal, and the two halves of the commutating inductance L1 and L2 are equal and tightly coupled by being wound on the same core. For purposes of analysis one half-cycle of operation is divided into five intervals, A, B, C, D, and E. The sequence of events taking place during the commutation process is shown in Fig. 6-9(b) for one phase of a three-phase bridge inverter supplying an inductive load.

INTERVAL A

It is assumed that SCR1 is conducting. Since the load is inductive, the rate of change of load current I_L will be small, and as a result the induced voltage across L1 will be small. If the device voltage drop is neglected, the voltage at point A will be equal to the positive busbar voltage. At the same time, the voltage across capacitor C1 will be zero, and the voltage across capacitor C2 is equal to the supply voltage V_d.

INTERVAL B

When SCR2 is turned on, the voltage across C2 is applied across inductor L2 and will in turn appear across L1. In other words, point Y is at the negative busbar potential, and point X is at a potential of $2V_d$ with respect to the negative busbar. SCR1 is reverse-biased, and turn-off is initiated. Since the emf in the core of the inductors L1 and L2 cannot change instantly, the load current I_L previously carried by L1 will be transferred immediately to L2. The load current i_L in the load is maintained by the capacitor currents i_{c1} and i_{c2}. During this period, capacitor C2 is discharging and thus supports the voltage across L2 and the induced voltage across L1; also capacitor C1 is charging to the dc source potential V_d. In the course of the charging cycle it will exceed the reverse bias across SCR1 produced by the voltage across L1, and SCR1 will be in a forward blocking state. During this same interval of time, the induced voltage across L2 will be carrying the oscillatory current in the L2–C2 circuit and the current through SCR2, I_{SCR2}, will increase to I_m at the point when C2 is fully discharged.

INTERVAL C

Capacitor C1 is now fully charged. The inductive load current i_L flows through diode D2, and point A is clamped to the negative busbar. Any stored energy remaining in L2 is dissipated as an I^2R loss through L2, SCR2, and D2.

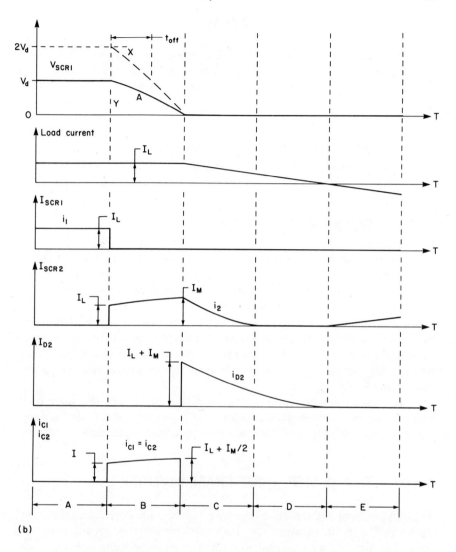

FIG. 6-9 *(cont.)* (b) voltage and current waveforms.

INTERVAL D

During this interval the current through L2 and SCR2 has decayed to zero, but because of the inductive nature of the load current, load current will still flow through D2.

INTERVAL E

The load current flow through D2 decreases to zero and D2 becomes reverse-biased; the load current is transferred to SCR2, if it is assumed that it is gated on. At this point the transfer of load current from SCR1 to SCR2 is completed. The next half-cycle would be then initiated by gating SCR1, and the process would then be repeated.

This commutation technique is known as the McMurray-Bedford circuit and has the disadvantage that under conditions of light load the commutating losses are high and, therefore, the inverter efficiency is low under light load conditions.

6-9   FREQUENCY CONTROL

Variable-speed ac motor control by the use of static inverters requires precise speed regulation over a wide range of frequencies. The output frequency of a static inverter is controlled by the rate at which the thyristors are gated on and off. The desired output frequency is controlled by a low-power master oscillator that generates a train of timing pulses, which, by means of logic control, can be directed to control the gating and turn-off circuits. Normally with ac motor control, the frequency control system is operated as an open-loop system, and will provide satisfactory results even if there are variations in the ac source voltage and frequency.

The most common type of master oscillator is the UJT relaxation oscillator, which can provide accuracies of ± 0.05 percent for any set frequency under normal operating conditions as met in industrial applications.

6-10   THE SIX-STEP, THREE-PHASE INVERTER

The simplest form of three-phase inverter is the three-phase bridge configuration shown in Fig. 6-10, in which the firing and commutating circuits have been omitted to simplify the explanation. Some commutation techniques have been previously explained in Sec. 6-8 for single-phase circuits, and can be readily extended to three-phase applications.

The thyristors in Fig. 6-10 are numbered in the sequence in which they are fired to produce positive-phase sequence voltages V_{AB}, V_{BC}, and V_{CA} at the output terminals A, B, and C. The output of the inverter can be obtained by either of the following gating firing sequences:

1. Three thyristors in conduction at the same time.
2. Two thyristors in conduction at the same time.

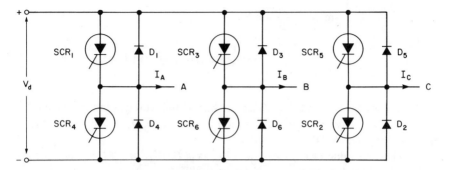

FIG. 6-10 Basic six-step, three-phase inverter.

In either case, gating signals are applied and removed every 60 deg of the output waveform, and as a result there are six distinct steps in every cycle.

6-10-1 Three Thyristors in Conduction at the Same Time

Gating the thyristors of Fig. 6-10 in the sequence SCR1, SCR2, SCR3, SCR4, SCR5, and SCR6 every cycle and leaving each in conduction for 180 deg of the output cycle will produce voltages, with respect to the negative busbar, V_{AN}, V_{BN}, and V_{CN} at the output terminals A, B, and C. The line-to-line voltages V_{AB}, V_{BC}, and V_{CA} are obtained by subtraction as follows: $V_{AB} = V_{AN} - V_{BN}$, $V_{BC} = V_{BN} - V_{CN}$, and $V_{CA} = V_{CN} - V_{AN}$, and are displaced 120 deg from each other. (See Fig. 6-11.)

The line-to-line voltages can be shown by Fourier analysis to be

$$V_{AB} = \sum_{n=1,3,5}^{\infty} \frac{4V_d}{n\pi} \cos \frac{n\pi}{6} \sin n\left(\omega t + \frac{\pi}{6}\right) \text{ volts} \qquad (6\text{-}22)$$

$$V_{BC} = \sum_{n=1,3,5}^{\infty} \frac{4V_d}{n\pi} \cos \frac{n\pi}{6} \sin n\left(\omega t - \frac{\pi}{2}\right) \text{ volts} \qquad (6\text{-}23)$$

$$V_{CA} = \sum_{n=1,3,5}^{\infty} \frac{4V_d}{n\pi} \cos \frac{n\pi}{6} \sin n\left(\omega t + \frac{5\pi}{6}\right) \text{ volts} \qquad (6\text{-}24)$$

Considering V_{AB}, we can simplify Eq. (6-22) to

$$V_{AB} = \frac{2\sqrt{3}}{\pi} V_d \left[\sin \omega t - \frac{1}{5} \sin 5\omega t \right.$$

$$\left. - \frac{1}{7} \sin 7\omega t + \frac{1}{11} \sin 11\omega t + \ldots \right] \qquad (6\text{-}25)$$

and similarly for V_{BC} and V_{CA}. Since the line-to-line voltages are the differences of two square waves, the triplen harmonics ($n = 3, 6, 9, \ldots$ etc.) in Eqs. (6-22), (6-23), and (6-24) are all zero.

The rms value of V_{AB}, V_{BC}, and V_{CA} is $\sqrt{(2/3)}V_d$, or $0.816V_d$, and the rms value of the fundamental is $\sqrt{6}\,V_d/\pi$, or $0.78V_d$.

If it is assumed that the load connected to the output terminals A, B, and C of Fig. 6-10 is delta-connected, then the load phase currents can be calculated from Eqs. (6-22), (6-23), and (6-24).

If the connected load is star or wye-connected, then the line-to-neutral voltages can be developed as shown in Fig. 6-12 and 6-13 and the load phase voltage is defined as

$$V_{AO} = \frac{3}{\pi} V_d \left[\sin \omega t + \frac{1}{5} \sin 5\omega t + \frac{1}{7} \sin 7\omega t \right.$$

$$\left. + \frac{1}{11} \sin 11\omega t + \ldots \right] \qquad (6\text{-}26)$$

and similarly for V_{BO} and V_{CO}, and thus the load phase currents can be calculated.

For either a delta- or star-connected load a six-step waveform is produced · as a result of applying six evenly spaced gating and commutating signals.

6-10-2 Two Thyristors in Conduction at the Same Time

From Fig. 6-11, it can be seen that SCR1 conducts from 0 deg to 180 deg, and then SCR4 conducts from 180 deg to 360 deg. This condition assumes that SCR1 is commutated instantly prior to SCR4 being turned on; if there is any delay in the turn-off of SCR1, a short circuit or "shoot through" will exist between the positive and negative busbars through SCRs 1 and 4. It can be seen that this arrangement of gating the SCRs is a possible cause of failure. The possibility of a "shoot through" condition can be reduced by arranging for a 60-deg delay prior to turning on SCR4 after SCR1 has had commutation initiated. Six commutations are still required per cycle of inverter output voltage. The associated voltage waveforms and gating sequences are shown in Fig. 6-14.

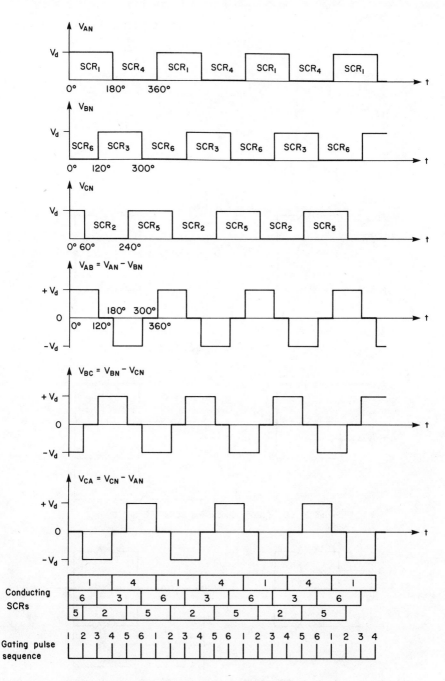

FIG. 6-11 Six-step, three-phase inverter, voltage waveforms, conducting thyristors, and gating sequence.

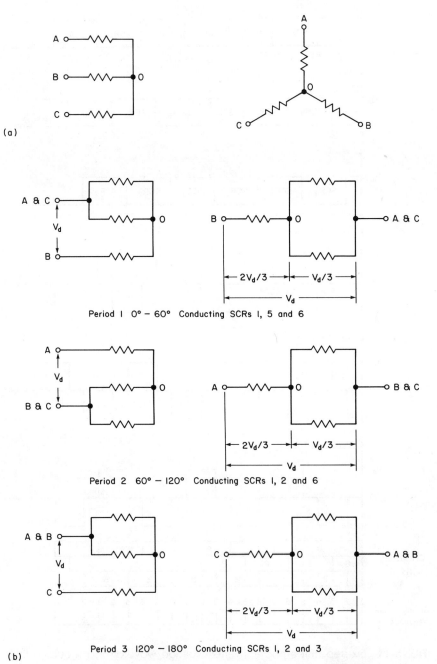

Period 1   0° − 60°   Conducting SCRs 1, 5 and 6

Period 2   60° − 120°   Conducting SCRs 1, 2 and 6

Period 3   120° − 180°   Conducting SCRs 1, 2 and 3

FIG. 6-12   Determination of line-to-neutral voltages for a star-connected load.
(a) Equivalent circuit; (b) phase voltages every 60° interval.

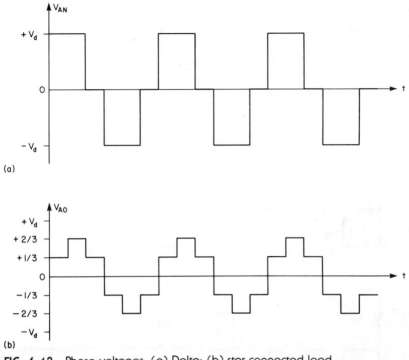

(a)

(b)

FIG. 6-13 Phase voltages. (a) Delta; (b) star-connected load.

6-10-3 Current Waveforms

To complete the analysis of a six-step, three-phase inverter, the relationship and nature of the phase and line currents supplying a typical balanced delta-connected inductive load are considered.

Since the voltage V_{AB} supplied to each phase is a square wave with voltage levels of $+V_d$, zero, and $-V_d$ [refer to Fig. 6-15(a)], the step changes of voltage will result in the load phase current rising and decaying in an exponential manner. The load phase currents i_A and i_B will be identical, but displaced 120 deg from each other. The line current drawn from the inverter is $I_A = i_C - i_A$. Similar load phase and line current waveforms can be developed for a star-connected load.

The currents in the SCRs and diodes can be developed from Figs. 6-15(e) and 6-10. At the instant that SCR1 is turned on and SCR4 is commutated off, I_A is negative; thus the line current must flow through D1 and in turn reduces the current drawn from the dc source. As long as $(i_C - i_A)$ is negative during the period of commutation, i_A will flow through D1 and SCR5 until i_C increases, making $(i_C - i_A)$ positive. At this point SCR1 conducts, provided there is a

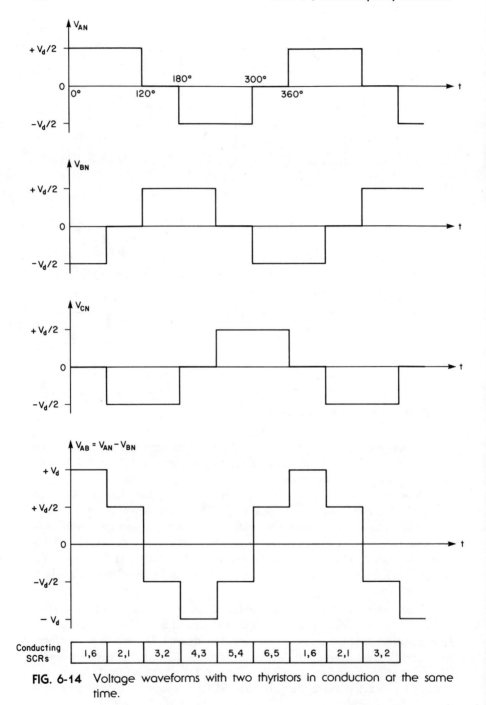

FIG. 6-14 Voltage waveforms with two thyristors in conduction at the same time.

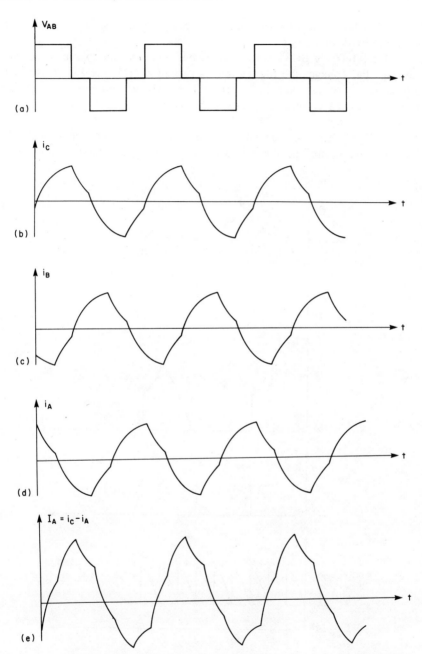

FIG. 6-15 Voltage and current waveforms for a six-step, three-phase inverter supplying a delta-connected inductive load. (a) Line voltage; (b), (c), and (d) load phase currents; (e) line current.

gating signal. Normally the gate signal is applied for 180 deg for inductive loads to ensure thyristor turn-on.

After 180 deg SCR1 is turned off, but since the line current is positive and lags the voltage, the line current is carried by D4 (see Fig. 6-16). The greater the lag of the current with respect to the voltage, the greater will be the magnitude of the current being carried by the thyristor at the instant of commutation.

In Fig. 6-15, the load phase currents consist of two components, the charging component and the discharge component. The charging component

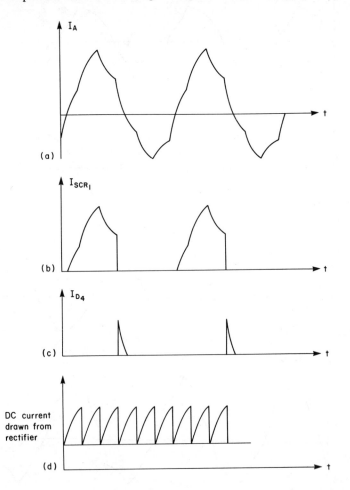

FIG. 6-16 Current distribution in a six-step inverter with an inductive load. (a) Line current waveform; (b) SCR1 current; (c) diode D4 current; (d) current drawn from rectifier.

flows through the dc source, and the discharge component circulates through a conducting SCR and diode. The dc current drawn from the dc source is the sum of the charging current components of the phase currents. When $(i_C - i_A)$ is negative during and just after the commutation period, current flows through D1 and reduces the current drawn from the dc source. During the interval 60 to 120 deg, SCR1 is the only thyristor connected to the positive dc busbar, and if $(i_C - i_A)$ is still negative, which would be the case for a low lagging power factor load, the dc supply current reverses, indicating a return of stored energy through the feedback diodes. Solid-state rectifiers will not permit the reversal of the dc current, and as a result a large capacitor is usually connected in parallel across the inverter input to prevent a rise in the dc voltage supplied to the inverter.

6-11   HARMONIC NEUTRALIZATION

The output of a dc link converter has a high harmonic content, since the converter is designed to operate over a wide frequency range, typically 10 to 200 Hz; hence it is not practical to attempt to design and apply filtering externally to the converter. There are several ways of minimizing the output harmonic content. One way is to use pulse width modulation techniques within the inverter. An alternative method is to combine a number of square-wave inverters, each of which is phase-shifted and fired at the desired output frequency. This method is known as *harmonic neutralization.*

Figure 6-17 shows a three-phase inverter, which consists of three single-phase bridge inverters. The transformer primaries that constitute the load of each converter have their secondaries connected in star and supply the phase load. This arrangement eliminates the third and triplen harmonics in the output waveform, although the higher harmonics remain. As can be seen from Fig. 6-18, a commutation occurs every 60 deg, resulting in six commutations per cycle, producing a six-step waveform.

It can be proved that there is a relationship between the number of commutations N per cycle of the output waveform and the lowest harmonic present. The relationship is that the lowest harmonics present are the $(N \pm 1)$ harmonics; for example, in the six-step output waveform, the lowest harmonics present are the fifth and seventh.

If the output voltage amplitude is varied by pulse width control, then the voltage applied to the primaries of the transformers can vary in width from 180 to 0 deg. Examples of the output waveforms for varying pulse widths are shown in Fig. 6-19.

A twelve-step output can be obtained by using two six-step inverters, with the gating signals of the second inverter being displaced 30 deg with respect to the first; that is, at 30, 150, and 270 deg. This combination results in twelve commutations per cycle of the output, and the lowest harmonics present will be

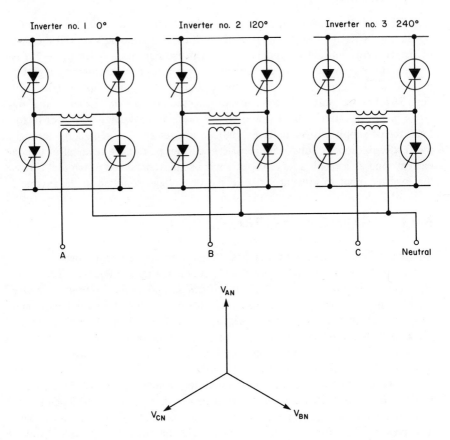

FIG. 6-17 A six-step inverter formed from three single-phase bridge inverters.

the eleventh and thirteenth. The amplitudes of the steps of the output waveform, by careful choice of the primary and secondary transformer winding ratios, can be made to closely approximate the average value of the stepped intervals of a sine wave, and thus the output waveform will be a very good approximation of a sine wave, even under minimum pulse width conditions.

If it is required to increase the kVA rating of an inverter, it is obvious that more benefit is gained by adding together simple inverters and offsetting them to create more steps to improve the output waveform, than is to be gained by paralleling the inverters. It must be appreciated, however, that the greater the number of steps, the greater the cost and complexity of the inverter, which results in the higher-step inverters being primarily used in high kVA applications.

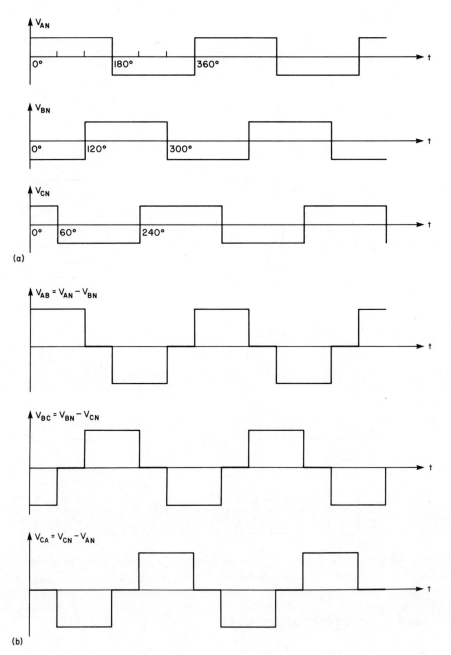

(a)

(b)

FIG. 6-18 Six-step inverter waveforms with 180° pulse width, 0° retard. (a) Line-to-neutral; (b) line-to-line.

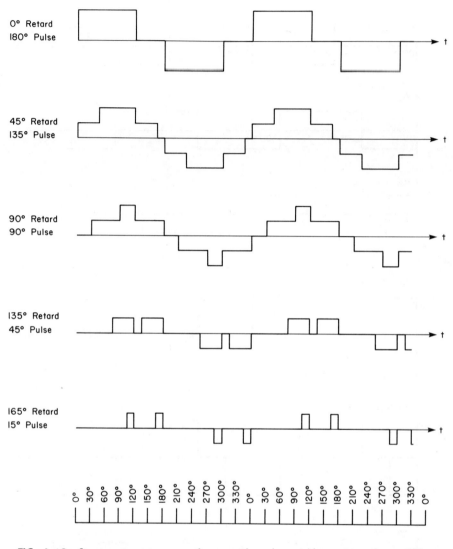

FIG. 6-19 Six-step inverter waveforms with pulse widths varying from 15° to 180°.

6-12   CYCLOCONVERTERS

The cycloconverter is a direct frequency changer that converts ac power at one frequency to ac power at another frequency by ac-to-ac conversion, without an intermediate dc link converter.

The concept of the cycloconverter is not new; it was originally developed in the 1930s and utilized grid-controlled mercury-arc rectifiers to reduce a standard three-phase, 50-Hz source to single-phase power at 16⅔ Hz for electric traction purposes by the German State Railways.

However, it was not until the introduction of the thyristor with its compact size, lower voltage drop, ability to be operated in any position, mechanical ruggedness, high-power and high-frequency switching capabilities, combined with logic and microprocessor control systems, that the cycloconverter re-emerged as a viable frequency changer.

The major applications of cycloconverters are low-speed, multimotor drives in the range up to 300 hp (224 kW) and large motor drives in the range from 2000 to 20,000 hp (1500 kW to 15,000 kW) with supply frequencies from 0 to 20 Hz. The system can provide reverse operation and regeneration, and the motor drive is started by reducing the stator voltage and frequency.

Another application that is used extensively in the aircraft industry is the variable-speed, constant frequency (VSCF) system, which provides, by a closed-loop control system, a regulated output voltage at constant frequency irrespective of speed changes in the engine-driven generator system.

Normally cycloconverters are operated in the frequency range of 0 to one-third of the supply frequency, in order to keep the harmonic content of the output voltage waveform within acceptable limits. However, it may be operated as a frequency multiplier, but the harmonic content is greatly increased and the converter efficiency is greatly reduced, although by the use of complicated rectifier circuits the output waveform may be improved.

Cycloconverters can produce a variable-frequency output by the use of phase-controlled converters or a fixed frequency output by envelope control.

The major advantages of using a cycloconverter are:

1. By the elimination of the dc link the overall converter efficiency is improved.
2. Voltage control is achieved within the converter.
3. Line commutation is obtained, with the total elimination of forced commutation components and circuitry.

6-12-1   The Single-phase to Single-phase Cycloconverter

The simplest way to understand the principle of operation of a cycloconverter is by studying the seldom used single-phase to single-phase cycloconverter.

The basic circuit shown in Fig. 6-20(a) consists of two two-pulse midpoint phase-controlled converters, one forming the positive group and the other the negative group, which is a dual converter configuration. The output current from each group flows only in one direction; therefore, in order to produce an

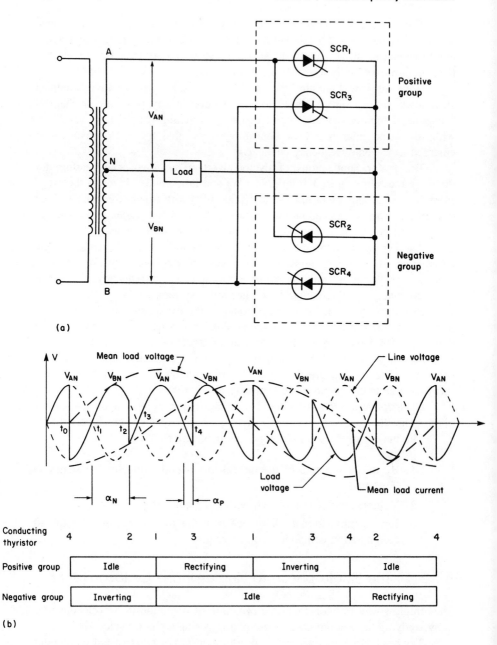

(a)

(b)

FIG. 6-20 Single-phase to single-phase two-pulse cycloconverter. (a) Basic circuit; (b) load voltage waveform and thyristor conducting sequence.

alternating current in the load, the positive and negative groups must be connected in inverse parallel. The positive group permits load current to flow during the positive half-cycle, and the negative group permits current flow during the negative half-cycle.

With reference to Fig. 6-20(b), it is assumed that the connected load is inductive, and that the secondary voltages V_{AN} and V_{BN} are antiphase as shown. At time t_0, line A is positive with respect to line B, and the load current is negative; SCR4 is conducting and remains in conduction because of the return of the stored energy in the reactive load, even though the anode-cathode became reverse-biased at time t_1. At time t_2, SCR2 is turned on and SCR4 turns off. When the load current becomes positive at t_3, SCR2 turns off and SCR1 is gated on and remains in conduction until time t_4, at which point SCR3 is gated on and SCR1 commutates off.

By varying the firing points of the thyristors, the amplitude $V_{do\alpha}$ of the mean output load voltage can be varied. Obviously, at time t_0 when the firing delay angle is 90 deg, then the output voltage is zero; it is a maximum when the firing delay is zero, and similarly for the negative half-cycle output voltage. Since the mean output voltage is

$$V_{do\alpha} = V_{do} \cos \alpha$$

the mean output voltage can be made to vary sinusoidally by suitably varying the firing delay angles α_P and α_N of the positive and negative groups so that the voltages will have the same amplitudes and frequency as each other at their output terminals. As a result, load current can flow in either direction irrespective of the polarity of the mean load voltage. From this, it can be seen that the direction of mean power flow can be into the load, the rectifying mode, or from the load to the ac source, the inverting mode. Regardless of the direction of power flow, the direction of current flow is such that positive load current is carried by the positive converter, and negative load current is carried by the negative converter.

As a result, each two-quadrant converter for any load other than unity power factor will be operating as a rectifier, or as an inverter, or will be idling and will contribute to the positive and negative half-cycles of the mean output load voltage [see Fig. 6-20(b)]. Consequently, the cycloconverter is capable of handling loads of any power factor from zero leading to zero lagging.

Since a firing delay of 90 deg results in a zero load voltage output, varying the firing delay angles about 90 deg in a sinusoidal manner at the desired output frequency will result in a mean output load voltage that is sinusoidal at the desired frequency. Voltage control is obtained by controlling the amount of variation of the sinusoidally modulated firing delay angle. The mean load current has been assumed to be sinusoidal, but in fact unless the load is highly inductive or filtering is applied, this will not be the case.

Normally the firing delay angles α_P and α_N of the positive and negative converters, respectively, are controlled simultaneously. In order to increase the mean output voltage of the positive converter, α_P is reduced, and α_N is increased. Similarly, to increase the mean output voltage of the negative group, α_N is reduced and α_P is increased. The firing delay angles are advanced and retarded with respect to the zero output voltage position, i.e., the 90 deg point, and α_P and α_N are related by

$$\alpha_P = 180° - \alpha_N$$

resulting in the output voltages of both converters being equal. However, the sum of the instantaneous voltages is not equal, and as a result a ripple voltage is produced which will drive current through both converters. The magnitude of this current must be limited by the use of a current-limiting reactor, the "circulating current mode of operation," or eliminated by blocking the firing signals to the converter that is not carrying load current, the "circulating current free mode of operation."

6-12-2   Three-phase Cycloconverter Arrangements

The output of the single-phase, two-pulse midpoint converter contains a high-amplitude ripple content; increasing the pulse number will cause a smaller amplitude ripple content in the load voltage waveform, as was the case with phase-controlled rectifiers.

The pulse number can be increased in a number of ways—for example, by using six three-pulse midpoint converters as shown in Fig. 6-21, connected to form three dual converters. This configuration permits only unidirectional operation, and requires 18 thyristors. A further increase in the pulse number can be obtained by using six six-pulse midpoint converters with interphase reactors (Fig. 6-22), requiring a total of 36 thyristors; once again, only unidirectional operation is permitted. The pulse number can be increased to twelve by using the circuit of Fig. 5-7.

The midpoint converter configurations all have a common output point to which the three-phase loads are connected. The bridge converter arrangement, on the other hand, requires the load to be connected across the common output terminals, thus requiring either the individual output phases or the three-phase inputs to the bridge converters to be isolated. The usual configuration is for the individual loads to be isolated (Fig. 6-23). Once again, this configuration requires 36 thyristors; however, four-quadrant operation is now obtainable.

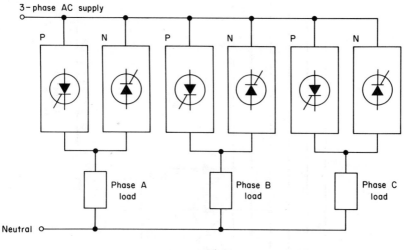

(a)

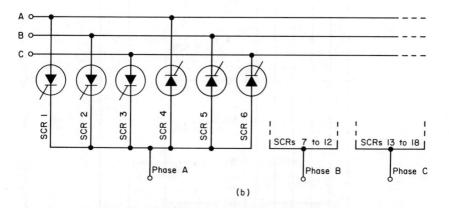

(b)

FIG. 6-21 Three-phase to three-phase three-pulse cycloconverter. (a) Schematic; (b) basic circuit.

6-13 CIRCULATING CURRENTS

As has been previously discussed, the output frequency is controlled by the rate of variation of the firing delay angles about 90 deg, and the amplitude of the mean output voltage is varied by the amount of variation of the firing delay angles about 90 deg, continuously maintaining the relationship $\alpha_N + \alpha_P = 180°$.

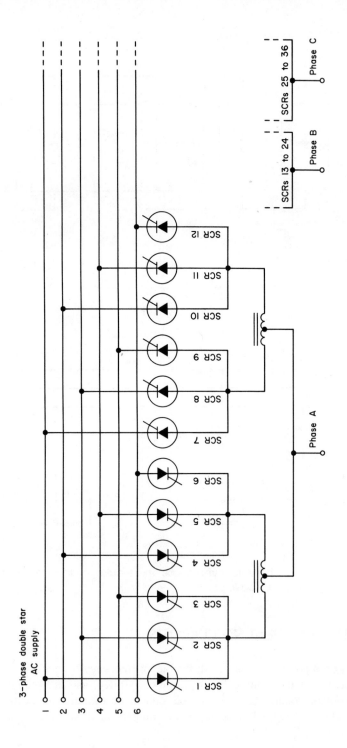

FIG. 6-22 Three-phase to three-phase six-pulse cycloconverter.

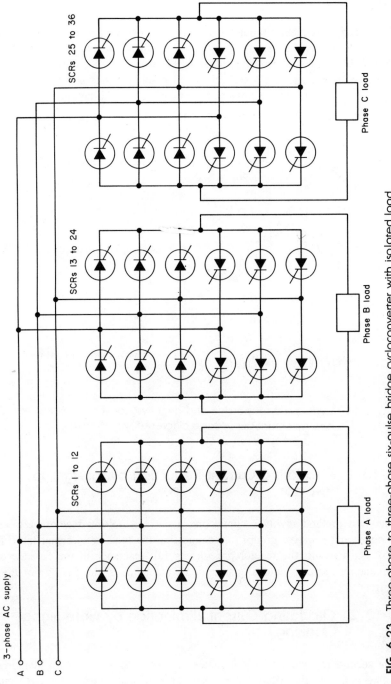

FIG. 6-23 Three-phase to three-phase six-pulse bridge cycloconverter with isolated load.

189

As a result, the mean output voltages of the positive and negative groups are equal. However, the instantaneous values of these two voltages are not equal (Fig. 6-24), and as a result it is possible for large harmonic currents to circulate which will increase the losses in the load circuit and increase the thyristor loading. There are two techniques by which the circulating current effects may be reduced.

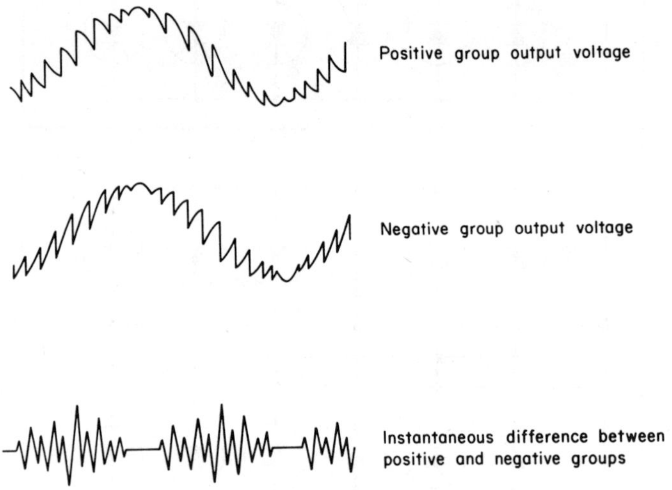

Positive group output voltage

Negative group output voltage

Instantaneous difference between positive and negative groups

FIG. 6-24   Cycloconverter positive and negative converter voltage waveforms and instantaneous voltage difference waveform.

6-13-1 Circulating Current Reduction by Intergroup Reactor

The first method involves connecting a current-limiting reactor between the positive and negative groups and connecting the load to its center tap (Fig. 6-25). As a result, the reactor presents its full reactance to the passage of the circulating currents, and a quarter of its reactance to the passage of the load currents.

6-13-2 Circulating Current Elimination by Gate Signal Blanking

The second method totally eliminates the circulating current by blanking the gating pulses to the inactive converter. This is accomplished by using current detectors in each phase.

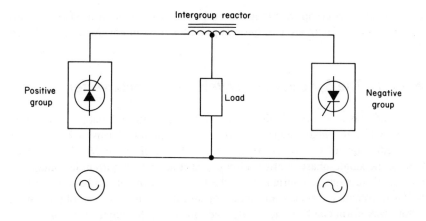

FIG. 6-25   Cycloconverter with intergroup reactor to reduce circulating currents.

This is the preferred method, but it is achieved by adding further complexity to the firing control circuitry. Under light load or unity power factor loads the mean output voltage distortion may be greater because of discontinuous conduction.

There are a number of other factors that affect the harmonic content of the output waveform. These are:

1. The pulse number of the converter; the higher the pulse number, the less the harmonic content.

2. The ratio of the source frequency to the output frequency; the lower the output frequency, the more closely will the output waveform approach the desired sinusoidal waveform, since the output waveform will be composed of more and more segments of the source waveform.

3. A high load reactance; the load current will be continuous and result in a reduction in the harmonic content.

4. The source reactance; the source reactance causes commutation overlap, which in turn increases the harmonic content.

6-14   ENVELOPE CYCLOCONVERTERS

The discussion up to this time has presented the application of cycloconverters for continuously variable-frequency applications. There are a number of applications where the output frequency is a fixed percentage of the source frequency or a limited number of fixed but lower frequencies. In these applications a cheaper and less complex configuration is obtained by the use of an envelope cycloconverter. There are two basic approaches used in envelope cycloconverters.

6-14-1 The Synchronous Envelope Cycloconverter

The single-phase configuration of Fig. 6-20 can, with a suitable logic control of the SCR gate signals, be made to produce output waveforms of 2:1, 3:1, 4:1, etc., reductions of the source frequency, as shown in Fig. 6-26. The disadvantage of the single-phase configuration is that it has a high harmonic content. By the use of three-phase configurations the harmonic content can be reduced [Fig. 6-26(c)]. When a variable-ratio input transformer is used, the amplitude of the output waveform can be varied as desired; however, the output waveform is still not a good approximation to a sine wave.

A better output waveform can be obtained when operating at one fixed output frequency, by the use of a six-pulse configuration fed by a star-double-

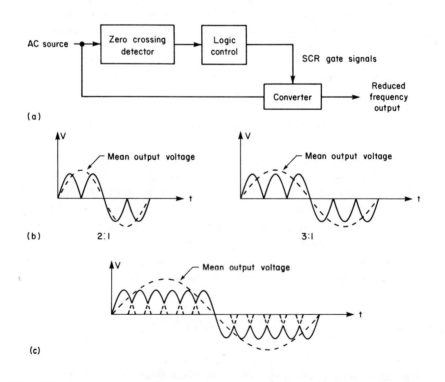

FIG. 6-26 Logic-controlled synchronous envelope cycloconverter. (a) Block diagram; (b) single-phase outputs; (c) three-phase output.

star transformer with four different secondary output voltages, which will then produce an output waveform that is a good approximation to a sine wave (Fig. 6-27).

With either of the above envelope cycloconverter configurations, the function of the thyristors is to ensure that the conducting group is turned off prior to the other group being turned on to prevent a short-circuit being produced across the ac source. In fact, these configurations operate satisfactorily with a resistive load, but with an inductive load they are incapable of regeneration. As a result, if they are to be used with an inductive load, a current detection sensor must be used and changes to the firing control circuitry must be made to ensure that firing signals to the incoming converter are blanked out until the load current goes to zero.

The basic disadvantage of the envelope cycloconverter is that it can operate only at a fixed frequency or at a number of discrete frequencies when controlled by a variable modulus counter. The major advantage, however, is a reduction in the complexity of the firing control circuitry.

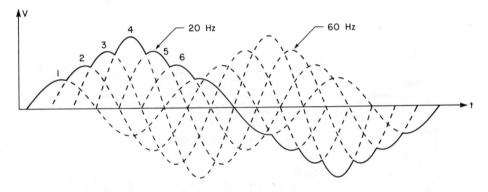

FIG. 6-27 Synchronous envelope cycloconverter, mean output voltage synthesized from the outputs of a six-pulse midpoint converter with varying amplitude voltages.

6-15   CYCLOCONVERTER FIRING CONTROL

The mean output voltage of a cycloconverter must vary from zero to a positive maximum to zero to a negative maximum and back to zero for each cycle of the output frequency. This is accomplished by biasing the firing delay angle to 90 deg and then oscillating the firing delay angle ±90 deg with respect to this point. As a result, for the positive half-cycle the positive converter firing delay angle is advanced, and the negative converter delay angle is retarded by the same amount. Similarly, during the negative half-cycle the firing delay angle

of the negative converter is advanced and that of the positive converter is retarded by the same amount so that $\alpha_P = 180° - \alpha_N$ at all times. Consequently,
the mean output voltages of both converters are equal and opposite in phase.

A commonly used method of generating the desired gate firing signals for
the thyristors is known as the *cosine-crossing* method. For a phase-controlled
converter, this method consists of comparing a cosine wave generated from and
synchronized to the ac source, with a variable dc reference signal by means of
a comparator, as shown in Fig. 6-28. It can be seen that as the amplitude of the
dc reference signal is varied the point of intersection with the cosine timing
curve will vary, and a firing signal can be generated corresponding to the point
of intersection, and as a result the firing delay angle can be varied as desired.
However, unless the dc reference signal is varying, the mean output voltage
$V_{do\alpha}$ will remain constant.

In the case of a cycloconverter the cosine-crossing method requires a
minor modification in order to produce a sinusoidal mean output voltage. Instead
of having a dc reference signal, a sinusoidal reference signal of the desired

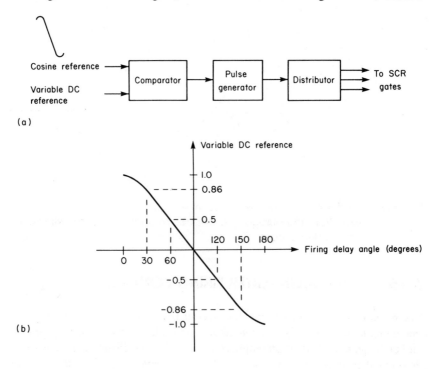

FIG. 6-28   Concept of cosine-crossing firing control. (a) Block diagram; (b) cosine
reference curve.

output frequency of the cycloconverter is compared to the cosine timing curves, and consequently the mean output waveform is modulated by the sinusoidal reference voltage. Voltage control is achieved by varying the amplitude of the reference voltage. If the cycloconverter is being operated in the circulating-current free mode, then current detectors are required to sense the current and produce blanking signals for the inactive converter. A block diagram of this control method is shown in Fig. 6-29.

There are a number of firing control techniques that may be applied to both phase-controlled converters and cycloconverters, such as the cosine-crossing method (also sometimes called the bias-cosine or ac-dc comparison method), the ramp crossing (also known as ramp and bias, or ramp and pedestal method), integral control, and phase-locked loop controls. In addition, considerable progress has been made in applying the microprocessor to the control of these devices These methods are dealt with in greater detail in Chapter 8.

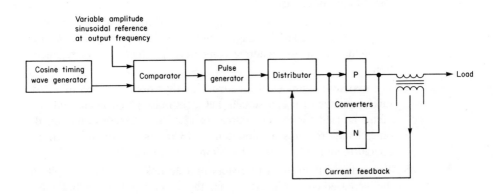

FIG. 6-29 Block diagram of a cycloconverter control system.

6-16   SUMMARY

At this point it would appear to be advantageous to summarize the relative merits and capabilities of the dc link converter and the cycloconverter.

In both cases the frequency of the output ac voltage is controlled by a reference oscillator system, operated as an open-loop control, and thus maintains a stable output frequency irrespective of load fluctuations. Voltage control and thus torque control are obtained in the dc link converter by pulse width modulation techniques and in the cycloconverter by controlling the amplitude of the reference signal. Also, a change of direction of rotation is accomplished by changing the firing sequence of the thyristors.

The differences between the dc link converter and the cycloconverter are:

1. The cycloconverter is an ac-to-ac frequency changer that is line-commutated and is used to produce a low frequency in one stage. The dc link converter, however, has two power conversions and requires forced commutation; as a result, the cycloconverter is more efficient.

2. The cycloconverter output frequency is usually limited to from zero to one-third of the source frequency in order to maintain the harmonic content of the output within acceptable limits. The reason that the harmonics are reduced at the lower frequencies is that the output waveform is composed of more segments of the source waveform than would be the case at higher frequencies. In the dc link converter pulse width or pulse width modulation techniques are used to reduce the amplitude of the harmonics, but the range of frequency variation is usually from 10 to 200 Hz with the variation being continuously variable in either case.

3. A bidirectional cycloconverter is capable of four-quadrant operation over its whole frequency range; however, unless the dc link converter is supplied by a two-quadrant converter, it is incapable of regeneration.

4. The maximum number of load-carrying thyristors in a three-phase dc link converter, if it is assumed that a six-pulse phase-controlled converter is used for the dc source, is 12. The cycloconverter, on the other hand, requires a minimum of 18 thyristors for unidirectional operation and 36 thyristors for bidirectional operation.

5. The reduced number of thyristors in a dc link converter, in spite of the increased cost of inverter-grade thyristors, combined with the less complicated control circuitry, gives it a competitive advantage compared to the cycloconverter, even though the lower duty cycle of the cycloconverter thyristors would reduce their cost.

6. A dc link converter supplied by a diode bridge presents a high input power factor, whereas the phase-controlled cycloconverter presents a low input power factor to the source, especially when operated at low output voltages.

In general, the cycloconverter is used in low-frequency, low-speed, high-horsepower applications, whereas the dc link converter may be used in all applications where wide frequency variations are required. With the recent introduction of high-current switching transistors, it is anticipated that the range of transistorized inverters will be extended into far higher horsepower (kW) ranges than are currently available.

REVIEW QUESTIONS

1. What are the advantages of selecting a variable-frequency ac drive as compared to a static dc drive in a variable-speed application?

2. In order to operate a polyphase induction motor from a variable-frequency source, what requirements must be met to provide a constant torque output? With the aid of torque-speed and torque-slip curves, discuss the advantages of constant torque control on starting torque, slip, and efficiency of a polyphase induction motor.

3. Discuss the advantages of using a static frequency converter versus a variable-frequency motor-alternator drive.

4. Explain the principle of four-quadrant operation of a polyphase induction motor.

5. With the aid of a schematic and waveforms show the production of a six-step output waveform by a three-phase inverter.

6. What are the advantages and disadvantages of controlling the output voltage of an inverter by varying the input dc voltage?

7. Discuss the control of the output voltage of an inverter by pulse width and pulse width modulation techniques and their effect upon the harmonics of the output waveform.

8. Explain with the aid of a schematic and waveforms the McMurray forced commutation technique. What are its advantages as compared to the McMurray-Bedford circuit?

9. Explain with the aid of a schematic and waveforms the McMurray-Bedford forced commutation technique. What are its disadvantages as compared to the McMurray circuit?

10. Discuss the principle of frequency control of a dc-link converter.

11. Discuss with the aid of a schematic and waveforms harmonic neutralization.

12. Explain with the aid of waveforms how the frequency and voltage of a phase-controlled cycloconverter are controlled.

13. What is meant by an envelope cycloconverter? What are its advantages and disadvantages, and how may its output waveform be improved?

14. Discuss with the aid of a block diagram the firing control of a cycloconverter supplying an inductive load.

15. Discuss the selection of a cycloconverter as compared to the dc-link converter in polyphase motor speed control applications.

16. Discuss the advantages and disadvantages of cycloconverters versus dc-link converters.

7 dc-dc Control

7-1 INTRODUCTION

As has been seen, ac phase-shift control and pulse burst modulation are used to control the amount of power transferred from an ac source to an ac load. Since the current source is alternating, the thyristors are automatically line-commutated every time the ac source voltage reverses.

The dc counterpart to ac phase-control techniques is the dc-dc converter or chopper, in which the mean load voltage supplied from a constant voltage dc source is alternately switched on and off by a thyristor (see Fig. 7-1).

The mean load voltage V_{do} can be varied in one of the following ways:

1. t_{ON} variable, t_{OFF} variable and the periodic time T constant; this is known as pulse width modulation.
2. t_{ON} constant, t_{OFF} variable; this is known as pulse rate modulation, frequency modulation, or variable mark-space control.
3. t_{ON} variable, t_{OFF} variable, which is a combination of pulse width modulation and pulse rate modulation.

As can be seen, it is possible to control the power supplied from a dc source to a load by switching the thyristors on and off. However, it is necessary to provide a means of applying reverse bias to the thyristor for a sufficient

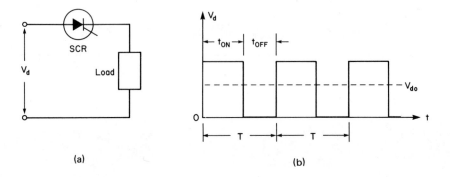

FIG. 7-1 DC-DC control, chopper control. (a) Basic circuit; (b) output voltage waveform.

duration so that the thyristor will turn off. This technique is known as *forced commutation*. Choppers have been applied to variable-speed dc drives, where they are used to provide a variable armature voltage for the speed control of separately excited dc motors, or a variable supply voltage for series motor speed control. The major advantage of choppers, as compared to the traditional electromechanical methods, is the higher efficiencies obtained by the elimination of the wasted energy inherent in the use of starting and control resistances.

7-2   FORCED-COMMUTATION TECHNIQUES

Forced-commutation circuits usually derive the reverse-bias voltage for turn-off from a charged capacitor, the only exception to this being where the reverse-bias voltage is initiated by an external pulse. Chopper forced-commutation circuits can be classified as follows:

1. Series-capacitor commutation.
2. Parallel-capacitor commutation.
3. Parallel capacitor-inductor commutation.
4. External pulse commutation.

For successful commutation, the commutation circuit must meet the following conditions:

1. The thyristor current must be reduced to zero.
2. The reverse-bias voltage must be applied to the thyristor for a period of time greater than the thyristor turn-off time.

3. The critical rate of reapplied forward voltage must not be exceeded.
4. In the case of inductive loads, the stored energy being returned to the circuit by the collapsing magnetic field must be diverted away by some means, e.g., a freewheel diode across the inductive load.

In addition, in order that the mean load voltage may be varied, the capabilities of the chopper circuit must be compared against the following criteria:

1. Is it capable of being operated in either the pulse width modulation or pulse rate modulation mode?
2. What is the range of voltage variation?
3. Does the commutation voltage provided by the commutation capacitor depend upon the load current?
4. Does the capacitor current cause the thyristor current rating to be exceeded?
5. What are the effects of failing to achieve turn-off?

7-2-1   Series-capacitor Commutation

In its simplest form the series-capacitor commutation circuit is shown in Fig. 7-2(a). When SCR1 is turned on, the thyristor carries only the capacitor charging current, which will decay to less than the holding current when the capacitor is charged to the source voltage V_d. With an underdamped resonating inductive load, the voltage on the capacitor will be in excess of V_d, and this reverse bias will assist the thyristor turn-off. With the circuit of Fig. 7-2(a), only one pulse of current through the load is possible unless the capacitor is discharged. Since the function of a dc chopper is to provide an unidirectional current through the load, there are two possible ways of achieving the discharge of the capacitor. In Fig. 7-2(b) a resistor is connected in parallel with the capacitor, the minimum time before applying a gate pulse to the thyristor being determined by the time constant of the R-C combination. An improved method of discharging the capacitor is shown in Fig. 7-2(c), where an inductor and a second thyristor are connected in parallel with the capacitor. The sequence of operation then is to turn on SCR1, which in turn charges C up to some voltage in excess of V_d; some time later and greater than t_{OFF}, SCR2 is turned on, and the capacitor is discharged through L and SCR2.

In series-capacitor commutation the load circuit forms part of the tuned circuit, which limits the control range, and any variations in load impedance reduce the effectiveness of the commutation circuit. This method of commutation is infrequently found in chopper applications, but it is used sometimes in inverter applications.

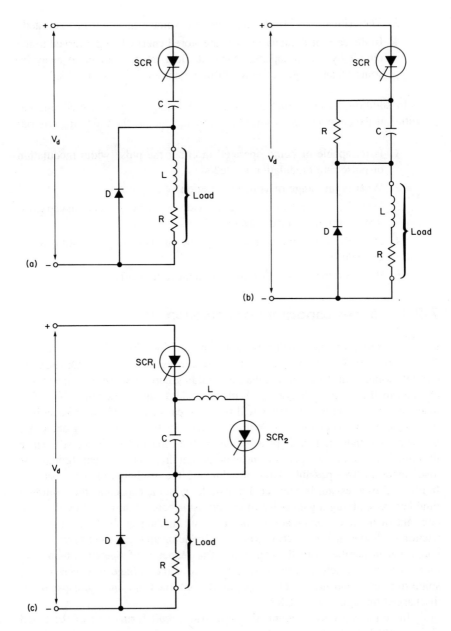

FIG. 7-2 Series capacitor commutation. (a) Basic circuit; (b) with capacitor discharge by a shunt resistor; (c) with capacitor discharge by a shunt thyristor.

7-2-2   Parallel-capacitor Commutation

In the parallel-capacitor circuit shown in Fig. 7-3(a), when SCR1 is turned on, the capacitor C will charge to potential V_d via R1, with its right-hand plate positive. Commutation is initiated when SCR2 is turned on, which will apply the positive potential of the capacitor to the cathode of SCR1 and turn it off. The period of time during which SCR1 is reverse-biased is known as the circuit recovery time t_{fr}, and it must be greater than t_{off} of the thyristor. The value of the capacitor C is obtained from

$$C = (I_L t_{fr})/V_d, \, t_{fr} > t_{off} \qquad (7\text{-}1)$$

where C = commutating capacitance, μF
 I_L = load current, amperes
 t_{fr} = circuit recovery time, μsec
 V_d = dc source voltage, volts

The major advantage of the configuration shown in Fig. 7-3(a) is that the capacitor charging current does not flow through the load. The disadvantage is that the potential on the capacitor is limited to the source voltage, and in high-current and high-frequency applications it may not achieve turn-off of the load-carrying SCR.

This circuit can be modified to achieve resonant charging of the capacitor by the addition of an inductor and a third SCR, as shown in Fig. 7-3(b). SCR1 and SCR3 are turned on simultaneously, and C is charged with its right-hand plate positive to a potential V_C greater than V_d, through L1 and SCR3 by resonant charging. The value of the capacitor C is obtained from

$$C = (I_L t_{fr})/V_C, \, t_{fr} > t_{off} \qquad (7\text{-}2)$$

where V_C = stored voltage on C, volts

If the capacitor C was discharged before SCR1 and SCR3 were turned on, then $V_C = 2V_d$.

This circuit has the following features:

1. Operation in either the pulse width modulation or pulse rate modulation modes is obtainable.

2. The minimum on time t_{ON} obtainable is equal to $\pi\sqrt{L_1 C}$, and this defines the minimum load voltage.

3. The peak charging current through L_1, SCR3 and SCR1, may require an inductance L1 that can safely handle the current.

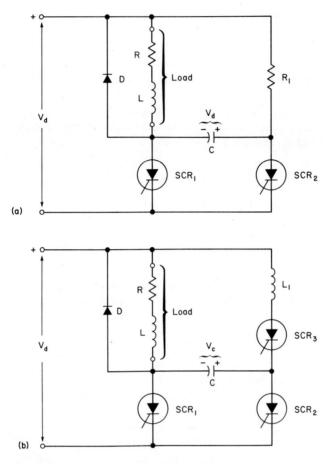

FIG. 7-3 Parallel-capacitor commutation. (a) Basic circuit; (b) with *L-C* resonant charging.

4. The minimum off time t_{OFF} is load-current-dependent, and this defines the maximum load voltage.

5. The maximum voltage that can be achieved on the left-hand plate of C is limited by the freewheel diode D to a maximum of V_d.

6. Since the capacitor charging current is superimposed on the load current through SCR1, the current rating of SCR1 must be increased accordingly.

7. In the event that SCR1 is not commutated off when SCR2 is turned on, then after the capacitor C has discharged SCR2 turns off; when SCR1 and SCR3 are gated on, C will recharge ready for the next commutation cycle.

8. A fault condition can exist with a loss of the devices, if SCRs 2 and 3 are on at the same time.

An alternative parallel-capacitor configuration, in which the capacitor charging current flows through the load, is shown in Fig. 7-4.

The capacitor C is precharged with the polarity shown to potential V_d by turning on SCR2; SCR2 automatically commutates off when the capacitor is fully charged. When SCR1 is turned on current flows in two paths, the load current I_L through the load, and the capacitor discharge current from C through SCR1, D1, and L1, which causes C to be resonantly charged with a reversed polarity to approximately $0.8V_d$. The reason for the low trapped charge on C is because of the resonant charging losses and the losses caused by SCR1 and D1. In addition, D1 must be chosen to have a low leakage current to prevent the charge being drained off C.

Commutation is initiated by turning SCR2 on, which reverse-biases SCR1 and reduces the current below the holding current, and the capacitor C recharges to its original condition through SCR2 and the load.

The features of this circuit are as follows:

1. Both pulse width modulation and pulse rate modulation modes of operation can be obtained.

2. The conditions noted for Fig. 7-3 apply to the minimum t_{ON} and t_{OFF}.

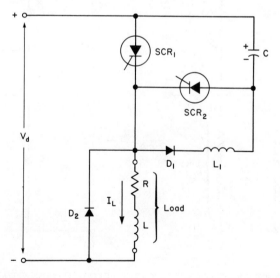

FIG. 7-4 Parallel-capacitor commutation with the capacitor charging current flowing through the load.

3. The commutating voltage on the capacitor is unaffected by load current.

4. The main disadvantage of this circuit is that if SCR1 fails to commutate when SCR2 is turned on, then the capacitor will discharge and SCR2 will turn off, thus making it impossible for C to recharge with the upper plate positive. To restore the commutation ability, the circuit must be isolated from the dc source and SCR2 must be regated after the power is restored.

5. The current rating of SCR1 must be increased to carry both the load and charging currents.

6. There is no possibility of a low-impedance fault path through the circuit, unless D2 has failed.

7-2-3   Parallel Capacitor-inductor Commutation

The basic parallel capacitor-inductor self-commutation circuit is shown in Fig. 7-5(a). In the case of series-capacitor commutation, the load circuit formed part of the tuned circuit, which limited the control range, and variations of load impedance reduced the effectiveness of the commutation circuit. The circuit

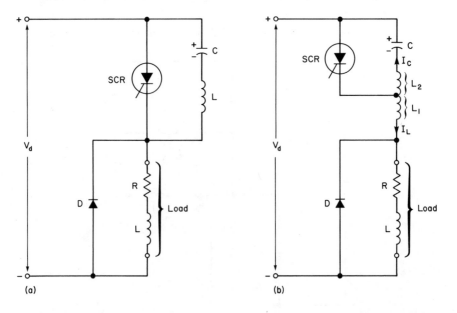

FIG. 7-5 Parallel capacitor-inductor commutation. (a) Basic circuit; (b) the Morgan circuit.

shown in Fig. 7-5(a) removes the problem by placing an underdamped L-C combination in parallel with the load-carrying SCR.

When power is applied from the dc source, the capacitor C charges up to approximately V_d with the polarity shown through C, L, and the load. When SCR1 is turned on, load current flows through the load, and simultaneously C discharges through SCR1 and resonantly charges up C with the capacitor polarity reversed, thus reverse-biasing SCR1. At the same time, the capacitor starts to discharge and reduces the current flow through SCR1 to less than the holding current and initiates turn-off. As soon as SCR1 is turned off, C will recharge with its original polarity through L and the load.

This circuit has the following properties:

1. It can be operated only in the pulse rate modulation mode.
2. The t_{ON} period is limited to $\pi\sqrt{LC}$; t_{OFF} is dependent upon the load resistance. The greater the load resistance, the larger will be t_{OFF}.
3. SCR1 can be fired only after C has fully recharged, or the circuit will fail to commutate.
4. Because of the restrictions on t_{ON} and t_{OFF}, the range of load voltage control is limited.

An improved version of this circuit is shown in Fig. 7-5(b). This circuit is known as the *Morgan Circuit*. This circuit depends upon the properties of a saturable reactor. When the reactor core is unsaturated, the inductance of the reactor is high; when the reactor core is saturated, the inductance of the reactor is low. With respect to Fig. 7-5(b), before SCR1 is turned on, C has been charged to V_d with the polarity shown, and since there is no current flow through the tapped reactor the core is desaturated and the reactor is in a high-inductance state. When SCR1 is fired the capacitor charging current I_C and the load current I_L will flow in opposing directions through the reactor, which then acts to keep the two currents building equally in a load-sharing manner. In effect, the reactor does not oppose the rise of current to the full load value except to keep the rates of rise equal, and the rate of rise will be determined mainly by the load inductance.

As soon as full load current flows, I_C will decrease and I_L will remain constant, and the voltage induced across $L2$, while very small, will be reversed in polarity. It will build in value as the charging current decreases. The decrease of I_C is determined by the resonant circuit $L_{UN} \cdot C$, which since the reactor core is still unsaturated will be high, thus giving a long period of oscillation. As the capacitor charging current I_C approaches zero, the net magnetizing current in the reactor approaches I_L, and the reactor core saturates When it does, the $L2$ inductance becomes very small and the full remaining voltage (reversed polarity) is applied to SCR1 to turn it off. The load current is then shunted into C, which then recharges to V_d. As I_L approaches zero, the reactor core desaturates and the cycle can begin again when SCR1 is turned on.

To ensure commutation, the load current I_L must be much greater than the core saturation current. Otherwise, saturation will not occur, and the SCR will not be reverse-biased by the capacitor. Also, saturation must take place before I_C reverses, or some of the capacitor voltage will be lost. The duration of the conduction period of the SCR is equal to 90 degrees of $L_{UN} \cdot C$.

Some of the features of the Morgan circuit are:

1. It is suitable only for pulse rate modulation control.
2. The commutating circuit time constants limit the duration of t_{ON}.
3. In the event of commutation failure, power must be removed from the circuit, and the process must be recycled.
4. The SCR is subjected to high initial di/dt, especially when the load is protected by a freewheel diode.
5. The load voltage regulation is very good.
6. This chopper is used very frequently for dc-dc control of dc motors and regulated dc power supplies.

Another L-C chopper configuration that overcomes the requirement that the capacitor be precharged prior to initiating the load current pulse is known as the *Jones circuit*, and is extensively used in electric vehicle control. The Jones circuit is shown in Fig. 7-6.

If it is assumed that C is discharged when SCR1 is fired, load current flows through $L1$ and the load, and since $L1$ and $L2$ are closely coupled, the capacitor C will be charged with the polarity shown by the induced voltage produced in $L2$ by the high di/dt in $L1$. The capacitor charge is trapped on C by the diode D1, until it is released by turning SCR2 on, at which point SCR1 is reversed-biased and turns off, and C charges with the polarity reversed via C, SCR2, $L1$, and the load.

Each time SCR1 turns on, C is charged. The magnitude of the charge depends upon the load current magnitude, with the result that under heavy load conditions, the charge on C will be at its greatest, which in turn aids commutation by decreasing the turn-off time of SCR1 under heavy loads.

7-2-4   External Pulse Commutation

This type of commutation depends upon the commutation energy being supplied from an external source. The conduction time of the load-carrying SCR, t_{ON}, is from the commencement of conduction to the initiation of turn-off. The duration of the commutation pulse must be equal to or greater than the turn-off time t_{OFF}

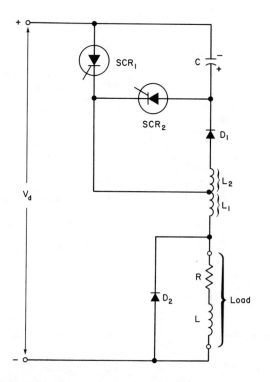

FIG. 7-6 The Jones circuit.

of the SCR. This permits operation of the chopper under pulse width modulation or pulse rate modulation control. Normally the reverse commutation voltage is applied either across the SCR or in series with it.

There are several configurations that may be used; they are illustrated in Fig. 7-7.

In Fig. 7-7(a), the commutating pulse to turn off SCR1 is provided by means of an auxiliary transistor switch $Q1$. The thyristor is assumed to be in conduction. When turn-off is desired, a signal is applied to the base of $Q1$, which reverse-biases SCR1, and the SCR is commutated off. The drive to the base of $Q1$ must be of sufficient duration to ensure turn-off and at the same time ensure that the transistor is driven into saturation. If $Q1$ comes out of saturation prior to turn-off being achieved, a commutation failure results. Accordingly, $Q1$ is selected to achieve turn-off under the heaviest load current and worst turn-off conditions of the SCR.

Figure 7-7(b) utilizes a pulse transformer with a square-loop B-H core to achieve turn-off. When SCR1 is conducting, the core of the pulse transformer is saturated by the load current in one direction. The application of a drive pulse

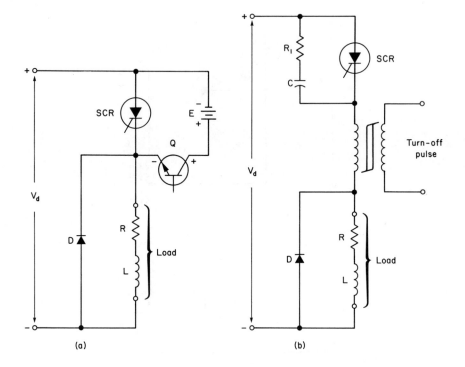

FIG. 7-7 External pulse commutation techniques. (a) By an auxiliary transistor; (b) by an external signal applied to a pulse transformer.

to the primary, in the opposite direction and larger in magnitude than the load current, will cause a reversal of the net magnetizing current for the pulse duration. When the net reverse magnetizing current exceeds a threshold value, the core flux reverses with a snap for approximately two or three microseconds. During this period a high reverse polarity voltage is developed across the transformer secondary, and this voltage pulse reverse-biases the SCR and it is turned off. With an inductive load, a freewheel diode must be connected across the load to prevent damage to the SCR by the induced load voltage spike as the load inductance discharges.

The major advantage to these external pulse commutation circuits is that both pulse width modulation and pulse rate modulation techniques can be used, and a commutation failure can be overcome by the external pulse source. In addition, commutation is independent of the load current and dc source voltage.

7-3 SUMMARY

The circuits that have been considered represent only a few of the many possible combinations. The most commonly used are the parallel-capacitor configurations, which give the greatest control flexibility over a wide range of dc source voltages. The parallel capacitor-inductor combinations are most applicable in high-voltage, low-current applications. The external pulse combinations, while more expensive to construct, offer the maximum control flexibility.

REVIEW QUESTIONS

1. What is meant by the term *chopper* or *dc-dc converter*, and what are the principles of the means of controlling the mean output voltage?
2. What are the commutating circuit requirements to achieve forced commutation of a chopper?
3. Discuss with the aid of a schematic, series-capacitor commutation. What are the disadvantages of this method of commutation?
4. Discuss with the aid of a schematic, parallel-capacitor commutation. What are the advantages and disadvantages of this circuit?
5. Explain with the aid of a schematic, parallel capacitor-inductor commutation. What are the advantages and disadvantages of this configuration?
6. Explain the operation of the Morgan circuit with the aid of a schematic.
7. Explain the Jones circuit with the aid of a schematic. List its advantages as compared to the Morgan circuit.
8. Describe the commutation of a dc chopper using an external pulse?

8 Firing Circuits

8-1   INTRODUCTION

The methods by which gating signals are produced and applied to the thyristor have been assumed in all the previous discussions. It must be recognized that the correct operation of ac static switches, ac phase control, controlled rectification, static inverters, cycloconverters, and dc-dc controls is dependent upon the provision of the firing signal to the thyristor at precisely the correct time. Equally as important, because of the relatively low voltage level of most thyristor gate signals, is the requirement that the gate circuit be free of all unwanted or spurious signals, since a noise signal of sufficient amplitude and duration will be enough to trigger a sensitive thyristor.

8-2   ELECTRICAL NOISE

Electrical noise consists of fast rise time electrical transients that appear in the supply and control wiring. In an industrial environment the sources of electrical noise are associated with high rates of change of voltage or current. The principal noise sources are switches, relay contacts, dc motor and generator commutator noise, inductive devices such as coils, solenoids and relays, induction heating, and thyristor switching. Noise can be transmitted by capacitive and inductive coupling, conduction, radiation, or common line injection. Normally the effects of electrical noise can be minimized by shielding, screening, isolating (both electrically and physically), filtering, and separating power wiring from the control wiring.

Some of the more common methods of suppressing noise are:

1. Noise suppression at its source. Usually it is not practical to try to eliminate noise, but minimization is a realistic objective. Some methods are to replace electromagnetic relays with solid-state relays with zero voltage switching capabilities. Capacitors connected across the relay contacts help to remove transients. Diodes, varactors, zeners, and *R-C* snubbers across inductive coils also help in the reduction of transient spikes.

2. Reducing the sensitivity of the control circuitry. In logic controls the use of high threshold logic (HTL) reduces the circuit sensitivity, but at the expense of speed capability.

3. Reducing the coupling between control circuits and the noise sources. In most situations involving noise suppression, this area yields the most effective results at a minimum cost. Some of the techniques that should be followed in all solid-state converter applications either during manufacture or during installation are:

   a. Provide the greatest possible physical distance between the control circuits and the power carrying circuits. This minimizes magnetic and capacitive coupling.

   b. Use the shortest distance between connections, and avoid grouping wiring runs.

   c. Magnetic coupling can be reduced, for example, by twisting conductors with a minimum of one twist for every two inches.

   d. Capacitive coupling can be minimized by using shielded cable, with the shielding being grounded at one point only.

   e. Never ground to the electrical common, but ground to the control system common.

   f. If the control system is supplied from the same ac source as the solid-state converter, the source should be filtered.

   g. In closed-loop control systems—for example, in current detection circuits—transformers operating at power line frequencies have a significant interwinding capacitance, and the screening between the windings should be grounded.

8-3 FIRING CIRCUIT PARAMETERS

In general, the following criteria should be met in all firing circuit designs:

1. The applied gate pulse must be of sufficient amplitude and duration so that the thyristor is turned on when required.
2. Provide accurate firing angle control over the required range.
3. Minimize the time delay between a change in the controlled variable— e.g., speed—and the corrective change in the firing delay angle.
4. Achieve a linear relationship between changes in the control signal and the converter output.
5. In polyphase applications accurately maintain the phase relationship between the firing signals, and ensure that there is simultaneous firing of all series- and/or parallel-connected thyristors in the same leg or phase.

8-4 FIRING DELAY ANGLE CONTROL TECHNIQUES

Numerous techniques of varying the firing delay angle α have been developed for three-phase systems; however, we intend to concentrate only on the commonly used approaches that form the basis for control in industrial applications. The basic concepts of cosine crossing, integral and phase-lock control will be examined.

8-4-1 Cosine Crossing

The cosine-crossing technique compares a variable dc reference voltage V_R to a cosine-timing signal so that under continuous conduction conditions there is a linear relationship between the dc reference voltage V_R and the mean dc output voltage $V_{do\alpha}$.

For a two-quadrant converter with phase control the mean dc output voltage is

$$V_{do\alpha} = V_{do} \cos \alpha \qquad (8\text{-}1)$$

and for a one-quadrant half-controlled converter with phase control the mean dc output voltage is

$$V_{do\alpha} = V_{do}(1 + \cos \alpha) \qquad (8\text{-}2)$$

The dc voltage ratio for a two-quadrant converter is

$$\frac{V_{do\alpha}}{V_{do}} = \cos \alpha \tag{8-3}$$

and for the one-quadrant converter is

$$\frac{V_{do\alpha}}{V_{do}} = (1 + \cos \alpha) \tag{8-4}$$

Equation (8-1), when plotted as shown in Fig. 8-1(a), is a cosine curve.

Assuming that the maximum value of the cosine timing wave is V_m and that the maximum value of the dc reference voltage V_R is equal to V_m, then we have

$$V_m \cos \alpha = V_R \tag{8-5}$$

and
$$\cos \alpha = \frac{V_R}{V_m} \tag{8-6}$$

This relationship is illustrated in Fig. 8-1(b), and from Eqs. (8-3) and (8-6)

$$\frac{V_{do\alpha}}{V_{do}} = \frac{V_R}{V_m} = \cos \alpha \tag{8-7}$$

and
$$\frac{V_R}{V_{do}} = \text{a constant} \tag{8-8}$$

The transfer function is linear. A similar relationship can be developed from Eqs. (8-4) and (8-6) for half-controlled converters by biasing V_R so that $V_R = 0$ when $\alpha = 180°$. Then

$$\frac{V_{do\alpha}}{V_{do}} = \frac{V_R}{V_m} = (1 + \cos \alpha)$$

and
$$\frac{V_R}{V_{do}} = \text{a constant} \tag{8-9}$$

The resulting transfer function is linear.

The basic concept of the cosine crossing method of controlling the firing delay angle is shown in block diagram form in Fig. 8-2.

The dc reference voltage V_R is the error signal derived by comparing a reference input signal, designating the desired speed, against a negative feedback

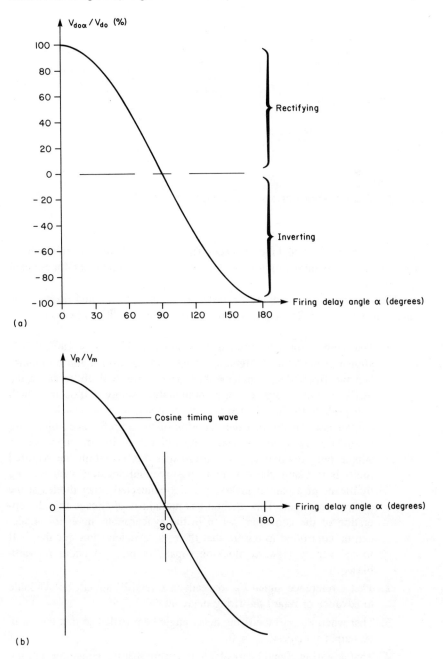

(a)

(b)

FIG. 8-1 Variation of $V_{do\alpha}/V_{do}$ vs. firing delay angle α for a two-quadrant converter.

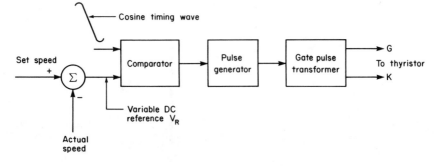

FIG. 8-2 Block diagram of cosine crossing firing control.

signal representing the actual motor speed. This signal is obtained from a voltage divider network sensing the motor counter emf or from a tachogenerator coupled to the motor.

The cosine crossing method, in order to accurately time and position the gating pulses to the SCRs, requires that the following conditions be met:

1. That control must be exerted to ensure that $0° < \alpha < 180°$; this is known as the "limit" signal or "end-stop" control. Failure to maintain the firing delay angle within these limits will either cause the SCRs not to be triggered, or a commutation failure will occur, which will probably cause device failure. A practical way to achieve this control is to ensure that the variation of the dc reference voltage V_R is slightly less than the peak values of the cosine timing wave. A simple but effective way to ensure that this is within the required limits is to clamp the maximum positive and negative values of V_R by means of a pair of inverse-parallel-connected zener diodes at the comparator input. This method has the added advantage that the operation of the converter, when in the synchronous inversion mode, can be controlled to ensure that there is sufficient time for the SCR to recover its forward blocking capability prior to being forward-biased.

2. That a reference signal V_R, or "advance-retard" signal, be available to advance or retard the firing delay angle.

3. That when $V_R = 0$ the firing delay angle be retarded so that the mean dc output voltage $V_{do\alpha} = 0$.

4. That a biasing signal be available to ensure that the converter will not operate in the rectifying mode irrespective of the firing delay angle when the converter is operating under current limit conditions.

These signal conditions are interpreted by the "gate pulse generator," which at the same time must provide a linear relationship between the reference signal and the mean output dc voltage $V_{do\alpha}$.

In three-phase converters the limit or end-stop signals are derived from the line-to-line ac source voltage, usually $L1$ - $L2$ as a reference transformed down to a suitable voltage level, phase-shifted and filtered by an RC network, and then applied to the gate pulse generator, so that it will permit gate pulses to be generated when it is greater than 0 volts.

Similarly, at the same time the cosine timing waves are also generated by phase-shifting techniques but leading the limit signals by 90 deg. The firing delay angle is obtained by comparing the positive or negative dc reference signal against the appropriate cosine timing wave by means of a comparator. As V_R is increased positively, the firing delay angle is reduced—that is, the firing point is advanced—and similarly, if V_R is increased negatively, the firing point is retarded.

The system in block diagram form is shown in Fig. 8-3. The output of the comparator initiates the production of a gate pulse signal, provided that the limit signal is permissive. The gate pulse signal is in turn amplified by a gate pulse amplifier before being applied to the primary of the gate pulse transformer. If optoelectronic couplers are used, the gate pulse amplifier may be eliminated, since these are compatible with TTL logic. Modern practice is to use NAND logic for the gate pulse generator. It is also common practice to provide a bias

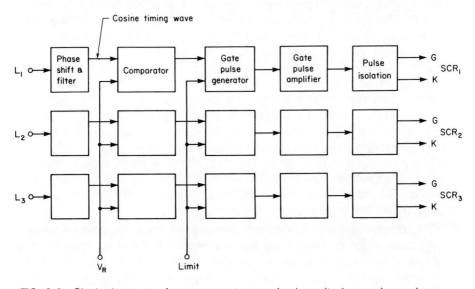

FIG. 8-3 Block diagram of cosine crossing method applied to a three-phase half-controlled converter.

signal that ensures that the firing delay angle is fully retarded when the reference input signal designating the desired speed is zero.

The major advantage of the cosine-crossing method of gate control is that it automatically responds to variations in the ac source voltage. A decrease in the source voltage will reduce the amplitude of the cosine timing wave and, since the dc reference signal is independent of this voltage and will remain constant, the firing delay angle will be reduced and the mean output dc voltage will remain constant. Similarly, with an increase in the ac source voltage the firing delay angle will be increased, with the result that once again the mean output dc voltage will remain constant.

The major disadvantages of the cosine crossing method are, first, since the cosine crossing wave is derived from the ac source voltage, any harmonics caused by system transients will be reproduced in the cosine timing wave with the possibility of the gate pulses being mistimed, thus producing a nonlinear transfer characteristic and even instability. Second, if there are variations in the ac source frequency the cosine relationship will vary, since the R-C phase shift network is frequency sensitive.

8-4-2   Integral Control

The two major disadvantages of the cosine-crossing method of firing control, namely, distortions of the ac source voltage waveform producing a nonlinear transfer characteristic and the phasing of the cosine signals being changed by ac source frequency variations, are overcome by the integral system of firing angle control. A block diagram illustrating the application of integral control to a three-pulse half-controlled bridge converter is shown in Fig. 8-4(a). The difference voltage between the reference voltage V_R and an attenuated feedback voltage from the converter output terminals is applied to the integrator and integrated [Fig. 8-4(b)]. The integrator output is zero at the desired firing points and has a frequency three times that of the ac source frequency. In turn, the integrator output is supplied to a comparator, which produces a train of pulses 120 deg apart at each zero point of the input. The comparator output is supplied in turn to a clock pulse generator, the output of which is applied to a three-stage ring counter and then is distributed by means of pulse transformers or optocouplers as gate trigger pulses to the thyristors.

If there is a decrease in the converter output voltage, the feedback voltage will decrease and the positive and negative integrals will also decrease; the zero points of the integrator output will advance. As a result, the trigger pulses to the thyristors will also advance, and the converter output voltage will increase. The system is then operating as a closed-loop control system responsive to variations in load demand.

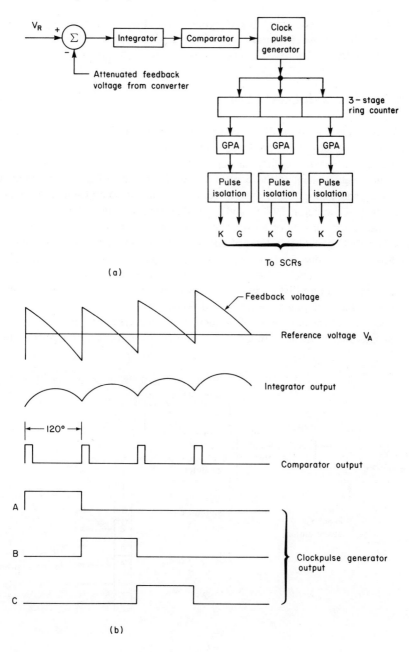

(a)

(b)

FIG. 8-4 Integral firing control. (a) Block diagram; (b) waveforms.

In addition, because of the integrator, variations in the supply voltage waveform, the feedback voltage waveform, and ac source frequency do not affect the system.

8-4-3   Phase-locked Loop Control

The phase-locked loop oscillator control of the production of thyristor gating pulses in converter control produces gate pulses at very accurate intervals and is not affected by voltage disturbances in the ac input to the converter.

The basic concept of the control system is shown in Fig. 8-5 as applied to a three-phase, six-pulse bridge converter. The major components are a voltage-controlled oscillator (VCO), and a six-stage ring counter. The voltage-controlled oscillator consists of an integrator, a comparator, and a reset or phase-lock. The VCO generates a train of pulses with a pulse repetition frequency f_1, which is directly controlled by a bias voltage E; in this case $f_1 = 6f$ where f is the ac source frequency. The VCO output is supplied to the six-stage ring counter, which in turn distributes the pulses in the correct sequence to the gates of the thyristors, the pulses being spaced 60 deg apart.

Because of the possibility of frequency drift the system is locked by means of a negative feedback loop in the VCO. End-stop control is also applied to the

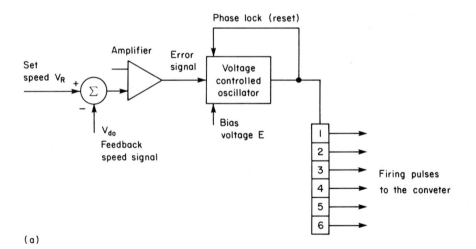

(a)

FIG. 8-5   The principle of phase-locked loop control. (a) Block diagram of system.

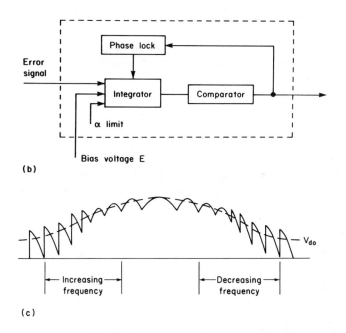

(b)

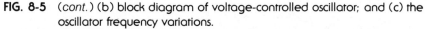

(c)

FIG. 8-5 (*cont.*) (b) block diagram of voltage-controlled oscillator; and (c) the oscillator frequency variations.

integrator to limit the amount of retardation of the firing delay angle [see Fig. 8-5(b)].

Any variations in the ac source frequency will cause a change in the firing delay angle and the mean dc voltage output of the converter. The resulting change in the error signal fed into the VCO will cause the oscillator frequency to change until it is once again $6f_2$, where f_2 is the new source frequency, but the firing delay angle and the mean output dc voltage will be returned to their original values. As a result, the VCO can accommodate quite wide frequency variations but still can maintain an accurate spacing of the firing pulses.

When operating as a closed-loop control system, if there is a decrease in speed, the feedback signal will decrease, and the increased error voltage applied to the VCO will cause the frequency of the output pulses to increase, which will result in the firing delay angle being advanced. In the case of an increase in the motor speed, the frequency of the VCO output will be decreased, and the firing delay angle will be retarded [Fig. 8-5(c)].

The phase-locked loop control system is a closed-loop system that provides very accurate response to changes in the controlled variable (motor-speed, voltage, current, etc.).

8-5   FIRING PULSES

The shape, amplitude, and duration of the firing pulse is determined by the gating requirements of the thyristor and the nature of the controlled load. For example in high *di/dt* applications, V_{GT} can range from 3 to 20 V and I_{GT} from as low as 0.25 to 4 A. At the same time, the rise time of the gate pulse should be less than one μsec, to ensure rapid spreading of the conducting area in the thyristor, and of sufficient duration to ensure that the latching current has been achieved. The pulse duration usually is as long as 50 μsec, although in some applications the duration may be as long as 200 μsec.

Pulses meeting these requirements may result in overheating of the gate-cathode junction. This effect can be minimized by a two-stage pulse (see Fig. 8-6).

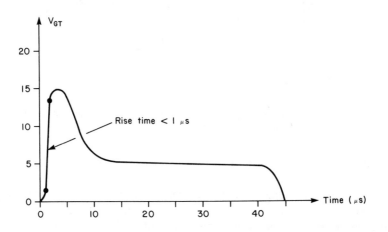

FIG. 8-6   Waveform of a two-stage firing pulse for high *di/dt* applications.

8-5-1   Double Pulsing and Long Pulses

In any full-controlled converter, when completion of the circuit depends upon two or more thyristors being in conduction at the same time and as conduction is transferred from one thyristor to the next as its anode voltage becomes positive, or during discontinuous conduction, one short pulse will not guarantee that all devices are triggered. One solution is to provide two pulses per cycle to each thyristor, the spacing of pulses being 2π/pulse number. In the case of a three-phase, full-controlled bridge converter the spacing is 60 deg [see Fig. 8-7(a)]. An alternative solution is to use long pulses. Long pulses are usually >60 deg [see Fig. 8-7(b)], but they still have the disadvantage that inductive

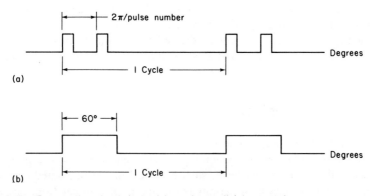

FIG. 8-7 Gate pulses for (a) double pulsing; (b) long pulses.

loads will delay the current buildup in the thyristor, so that at the end of the pulse a latching current has not been achieved. The only way to ensure that the thyristors are turned on is to supply a gating pulse of 120 deg duration, i.e., equal to the conduction period of the thyristor. There are two major disadvantages to this scheme: First, it is difficult to produce a pulse of 120 deg duration, and second, it increases the probability of overheating the gate-cathode junction.

8-6   PULSE ISOLATION

In any power converter, differences in potential between the various gates and between the individual thyristors and the gate pulse generators exist. Protection must be provided by electrical insulation or isolation between the thyristors and the gate pulse-generating circuits.

8-6-1   Pulse Transformers

A commonly used method of providing electrical insulation is by the use of a pulse transformer. Pulse transformers are available in a number of different mounting arrangements suitable for printed circuit board applications or dual-in-line packages containing two, three, or four transformers for integrated circuit applications.

Normally there is a primary winding and one or more isolated secondary windings, which permit simultaneous gating signals to be applied to thyristors in series and parallel configurations. These windings usually are tested at 2500 V to ensure that all windings are electrically isolated from each other.

The electrical requirements of a pulse transformer are, first, that the windings must be tightly coupled to minimize leakage inductance and thereby to ensure that the output pulse will have a fast rise time and, second, that the insulation be great enough to provide the isolation required. This latter requirement interferes with the first, and as a result the design must be an acceptable compromise.

8-6-2   Optocouplers

Optocouplers are a combination of optoelectronic devices, normally a light-emitting diode (LED) and a junction-type photoconductor such as a phototransistor assembled into one package, such as a dual-in-line integrated circuit arrangement. This arrangement permits the coupling of a signal from one electronic circuit to another and at the same time maintains an almost complete electrical isolation between the circuits.

The most commonly used optocouplers consist of an input light-emitting diode operating in the infrared or near-infrared region and an output silicon photodetector, such as a phototransistor, a photo-Darlington, or a photo-SCR (see Fig. 8-8).

When using a pulse transformer, gate trigger power is transferred from the trigger source to the thyristor gate and there is no leakage in the gate circuit. In

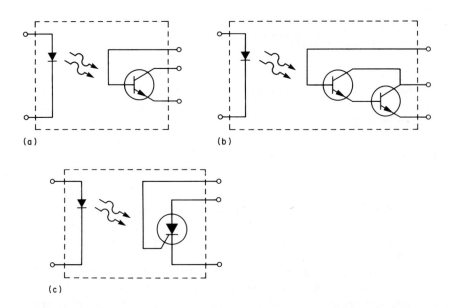

FIG. 8-8 Typical optocouplers. (a) Phototransistor; (b) photo Darlington; (c) photo SCR.

the case of an optocoupler, however, a power source is required on the load side. If this source is derived from the anode supply of the thyristor, the blocking voltage of the SCR cannot exceed that of the optocoupler. This disadvantage can be overcome by one of two means: first, by providing a separate power supply which provides greater flexibility at an added expense (more than one power supply will be required, if the load-carrying SCRs do not have common reference points), and second, by deriving the optocoupler power from a voltage divider across the input to the SCR, which will have the disadvantage of providing a leakage path around the SCR during the nonconducting period.

Typical operating speeds are 100 to 500 kHz for phototransistors, 2.5 to 10 kHz for photo-Darlingtons, and 2 to 20 μsec turn-on time for photo-SCRs. The type selected depends upon the power requirements of the gate circuit.

8-6-3   Output Stages

Probably the simplest application of the pulse transformer is shown in Fig. 8-9(a), where an NPN transistor is connected in series with the primary. When a signal is applied to the base of the transistor from the gate pulse generator in the firing control, the transistor saturates and the voltage V is applied across the transformer primary, and induces a voltage pulse at the secondary terminals which is applied to the SCR. When the drive pulse to the base of the transistor is removed, the transistor turns off, and the current caused by the collapsing magnetic field in the transformer flows through the diode D connected across the primary; the voltage across the primary momentarily reverses.

The core flux is unidirectional, and the core-magnetizing current as well as the pulse-forming current both have to be carried by the transistor. As a

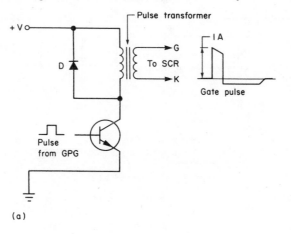

(a)

FIG. 8-9 Gate pulse isolation. (a) Simple pulse.

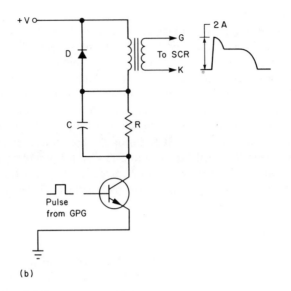

(b)

FIG. 8-9 *(cont.)* (b) two-stage pulse.

result, in order to minimize the core size, this type of configuration is usually
limited to the production of pulses between 25 to 50 μsec, and cannot be used
in long pulse applications.

In order to produce two-stage pulses for high *di/dt* applications, the circuit
is modified as shown in Fig. 8-9(b). The circuit operates as before, except that
the charging of the capacitor prolongs the duration of the overall pulse.

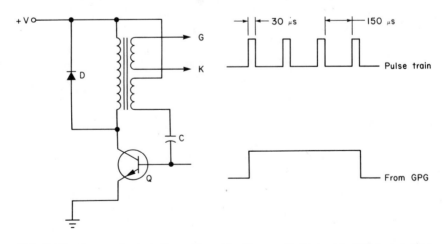

FIG. 8-10 Gate pulse isolation using a blocking oscillator to simulate long pulses.

Long pulses can be simulated by using a blocking oscillator configuration, as shown in Fig. 8-10. This application requires a pulse transformer with two secondaries and will produce a train of pulses as long as there is a base drive on the transistor by means of a positive feedback signal to the base of the transistor. The major disadvantage of this circuit in three-phase applications is the fact that, it is necessary to provide simultaneous pulses to two or more thyristors, but there is no guarantee that the thyristors will receive these pulses at exactly the same instant. A secondary disadvantage is that the regenerative circuit is very susceptible to noise and may cause inadvertent triggering of the thyristors.

A much more effective way of producing pulse trains for long pulsing is to use a common master oscillator and apply the output to the thyristors as dictated by the firing control circuit. This scheme is illustrated in Fig. 8-11,

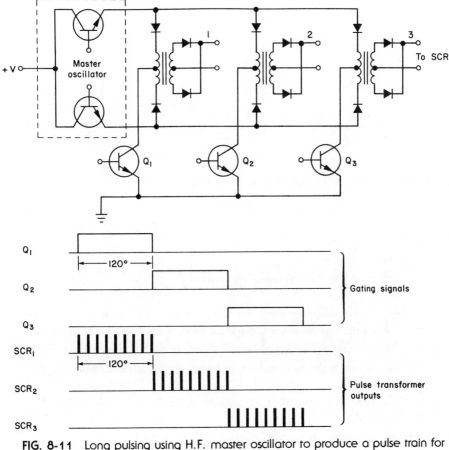

FIG. 8-11 Long pulsing using H.F. master oscillator to produce a pulse train for a three-pulse converter.

where the master oscillator produces a train of high-frequency pulses, typically between 5 to 20 kHz. The center-tapped pulse transformers are connected in parallel across the master oscillator output; base signals are applied to the gating transistors from the firing control circuit. The resulting pulse trains are applied to the thyristors for 120-deg conduction, and since the pulses are all derived from the master oscillator, the potential conducting thyristors will receive simultaneous gating signals.

Optocouplers provide an excellent means of coupling between TTL logic or a microprocessor output to thyristors, since the output currents may not be sufficient to cause the thyristor to latch. Figure 8-12 illustrates the use of a photo-Darlington optocoupler, which because of its higher current gain can be used for triggering a thyristor.

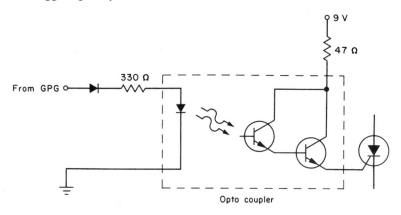

FIG. 8-12 Gate pulse isolation using an optocoupler.

REVIEW QUESTIONS

1. Explain what is meant by electrical noise. Discuss the causes and methods of noise suppression that are commonly used in solid-state drive systems.
2. Discuss the criteria that should be met by all firing circuit designs.
3. Discuss the cosine crossing method of firing control and the conditions that must be met.
4. Discuss the advantages and disadvantages of the cosine-crossing method of firing control.
5. Discuss the integral method of control of firing pulses.

6. Explain the principle of the phase-locked loop method of firing control. What are the advantages of this method?

7. What is meant by double pulsing and long pulses? Why are they required?

8. Why is pulse isolation necessary? How may it be achieved?

9. Discuss the requirements that must be met when using optocouplers for gate circuit isolation?

10. Explain with the aid of a schematic the production of pulse trains for control of a three-phase converter?

9 Rotating Machine Control

9-1 INTRODUCTION

The previous chapters have concentrated upon the development of the under-standing necessary to apply the thyristor to the control of dc and ac machines. It is assumed that the reader has a good working knowledge of the principles of operation and characteristics of dc and ac machines.

9-2 DC MOTORS

The most important property of the dc motor is that three different operating characteristics are obtainable by the method of field connection. The separately excited machine with the armature and field windings supplied from two separate sources permits speed control below base speed by supplying the armature with a variable voltage. Speeds above base speed are obtained by reducing the field current.

The shunt motor, in which the armature circuit is in parallel with the field circuit, gives an almost constant speed output over its operating load range. The series motor, in which the armature and field windings are connected in series to a dc source, has a characteristic in which the speed is inversely proportional to the torque. The compound motor, in which the armature is in parallel with the shunt field and a series field connected in series with this combination, has an operating characteristic that is in between that of the shunt and series motor, depending upon the degree of compounding given by the series field. The basic armature and field connections for these motors are shown in Fig. 9-1.

The performance of a dc motor is readily developed, based on the following assumptions:

1. The motor is operating under steady-state load conditions.
2. The magnetic circuit is not saturated.
3. The rotational losses can be neglected.
4. The armature resistance R_a is negligible, which is a valid assumption for motors above 5 hp (3.73 kW).

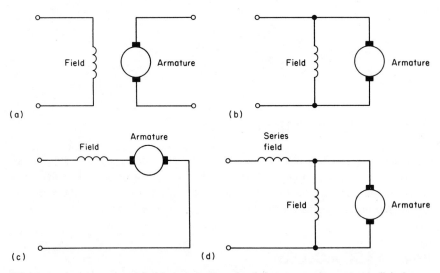

FIG. 9-1   Armature and field connections for (a) separately excited, (b) shunt-connected, (c) series-connected, (d) compound-connected dc motors.

9-2-1   Ideal Motor Characteristics

Consider the separately excited motor connected, as shown in Fig. 9-2, to a source of armature voltage V_a. Then, the armature circuit equation is

$$V_a = E_c + I_a R_a \tag{9-1}$$

If it is assumed that R_a is small enough to be neglected, which is a reasonable

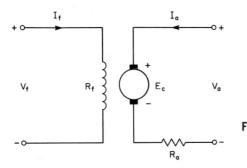

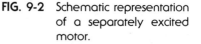

FIG. 9-2   Schematic representation of a separately excited motor.

assumption for large motors, then the armature voltage drop I_aR_a is also small and Eq. (9-1) can be rewritten as

$$V_a \simeq E_c \qquad (9\text{-}2)$$

The steady-state operation of the motor is determined by the fact that the generated or counter-emf E_c of the motor is slightly less than the applied armature voltage V_a. The counter-emf E_c is proportional to the speed of rotation of the armature and the field flux, and can be represented by

$$E_c = K\Phi S \qquad (9\text{-}3\text{E})$$

where $K = (ZP/60a) \times 10^{-8}$
         = constant of proportionality
        S = speed, rpm

or $$E_c = k\phi\omega \qquad (9\text{-}3\text{SI})$$

where $k = (ZP/2\pi a)$
         = constant of proportionality
        ω = angular velocity, radians/sec

Then, solving for speed, we have

$$S = \frac{E_c}{K\Phi} \simeq \frac{V_a}{K\Phi} \text{ rpm} \qquad (9\text{-}4\text{E})$$

or $$\omega = \frac{E_c}{k\phi} \simeq \frac{V_a}{k\phi} \text{ rad/sec} \qquad (9\text{-}4\text{SI})$$

Both these equations show that the speed of a dc motor is directly proportional to the counter-emf, and therefore to the applied armature voltage V_a, and inversely proportional to the field flux.

The torque developed by a dc motor is directly proportional to the armature current I_a and the flux per pole Φ (or ϕ); that is,

$$T = KI_a\Phi \text{ ft-lb} \tag{9-5E}$$

or

$$T = kI_a\phi \text{ N·m} \tag{9-5SI}$$

The horsepower (kW) output is proportional to both torque and speed; therefore,

$$\text{hp} = KTS \tag{9-6E}$$

or

$$\text{kW} = kT\omega \tag{9-6SI}$$

Equations (9-5) and (9-6) show that if the armature current I_a remains constant, then:

1. By varying the applied armature voltage V_a, with the flux per pole remaining constant, the output torque remains constant, and the output horsepower (kW) is directly proportional to the speed. This is known as the *constant torque mode*.
2. By applying rated armature voltage, and weakening the field flux per pole, the output horsepower (kW) remains constant and the output torque decreases. This is known as the *constant horsepower (kW) mode*.

The base speed of a dc motor is defined as the speed produced with rated armature voltage and rated field flux.

It can be seen that when the field flux is held constant at its rated value, variation of the applied armature voltage will result in variations of speed from zero to base speed, and the motor is said to be operating as a constant torque drive. When the applied armature voltage is at rated value, then reducing the field flux results in speeds above base speed being produced, and the motor is said to be operating as a constant horsepower (kW) drive. These conditions are illustrated in Fig. 9-3.

Since in the dc shunt motor the armature and field are in parallel, it is impractical to attempt speed control by means of variation of the applied armature voltage, although in the case of the separately excited motor and the series motor it is practical.

Usually, constant torque adjustable-speed drives are found in machine tool, conveyor, hoist, etc., applications and are limited by the capability of the solid-state control unit, not by the ac source power available. Constant horsepower (kW) drives are used in applications where a variable torque output is required, e.g., center-driven winders, electric traction motors, and cranes.

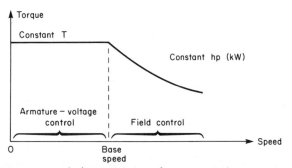

FIG. 9-3 Torque vs. speed characteristics of a separately excited motor with variable armature voltage and field control.

When a motor is operating at speeds above base speed by using field-weakening techniques, the range of speed control depends upon whether the motor is designed for fixed-speed operation or adjustable-speed operation. In the former, speed may be varied over a range of 1.5 to 1, and in the case of an adjustable-speed motor the speed control range can be of the order of 6:1. The limiting factors of the speed control range are both mechanical and electrical. The mechanical limitations are imposed by the centrifugal forces developed, and the electrical limitations are primarily the current that can be commutated, which becomes an increasing problem, since armature reaction effects increase as the field is weakened, and the time available for commutation decreases. In addition, the speed regulation of the motor worsens, and the output torque decreases. High-speed motors require special frames and stabilizing windings for stable operation, which add significantly to the motor cost.

9-3 CLOSED-LOOP PHASE-CONTROLLED DC MOTOR SPEED CONTROL SYSTEMS

The block diagram shown in Fig. 9-4 shows a typical closed-loop armature voltage dc motor speed control system. There are two requirements that must be met by the system: first, to limit the inrush starting currents, and second, to control the motor speed at whatever speed setting has been designated.

9-3-1 Starting

From Eq. (9-1), it can be shown that the armature current I_a at any instant is

$$I_a = \frac{V_a - E_c}{R_a} \tag{9-7}$$

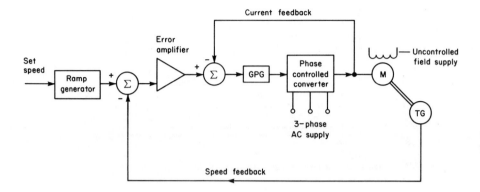

FIG. 9-4   Block diagram of a phase-controlled dc motor speed control system.

At the instant of starting, the counter-emf E_c is zero, and since in integral horsepower (kW) dc motors the armature resistance R_a is usually less than one ohm, it can be seen that if the voltage applied to the armature circuit is not reduced to a low value, the resulting armature current will be considerably in excess of the normal rated value. This high starting current would generate a high starting torque, but at the expense of damage to the machine. In traditional electromagnetic dc motor control, the inrush armature current at start is reduced by means of added resistance in series with the armature. Normally, inrush armature current is limited to 150 to 175 percent of the normal full-load current. Apart from the damage that may be caused to the machine, the maximum inrush current limits are also restricted by the utility authorities, in order to minimize system voltage disturbances to other customers. In a solid-state controller the inrush currents are limited by regulating the rate of advance of the firing delay angle by means of a ramp generator.

9-3-1-1   Ramp Generator

The function of a ramp generator is to provide a controlled acceleration of the dc motor when starting the drive, irrespective of the set speed desired. Usually the acceleration time is adjustable between two to ten seconds. A schematic of a typical ramp generator circuit is shown in Fig. 9-5. A dc reference input signal, typically 0 to 10 V, is applied to the inverting input of the operational amplifier $U1$, which is a high-gain inverter. The diodes, together with the potentiometers $P1$ and $P2$, form a variable gain amplifier; the potentiometers are used to control the acceleration and deceleration rates of the drive. The inverted output of $U1$ is applied to the inverting input of $U2$, which is connected as an

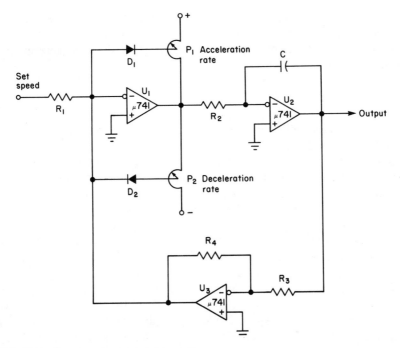

FIG. 9-5 Ramp generator schematic.

integrator whose output is a positive-going ramp under starting conditions, and is used to provide control of the firing point of the thyristors in the converter section. Also, a signal is taken off the output of $U2$ and applied to the inverting input of $U3$. The output of $U3$ will be a negative-going ramp. This signal is then compared at a summing junction with the reference input signal; when the two signals are equal, their sum is zero, and the output of the ramp generator will be equal to the reference input signal. In turn, any variations of $U2$ output will immediately be corrected by the negative feedback loop through $U3$.

As can be seen, the drive will start with the firing delay angle fully retarded, and therefore a minimum voltage is applied to the armature circuit. The rate of increase of the applied armature voltage is controlled by the slope of the ramp output at $U2$. As an added safeguard, a current limit or torque limit signal is also generated, so that if the rate of acceleration of the connected load produces currents in excess of a predetermined amount, a signal is generated which will cause the firing angle to be retarded and the output voltage of the converter to be reduced to bring the armature current back within the prescribed limits. The current limit signal can be obtained either by using a dc shunt in series with the armature circuit, which will produce a voltage proportional to the armature current, or by the use of current transformers in the ac supply lines to the converter. The secondary outputs of the current transformers are rectified, and once again a dc voltage signal proportional to the armature current is generated. The current feedback signal is compared against the dc signal from the error amplifier (see Fig. 9-4).

9-3-1-2   Speed Control

In order that the drive will operate at the desired speed, it is necessary to generate a dc voltage signal that accurately represents the actual motor speed. There are two basic methods: first, to sense the counter-emf of the armature circuit and reduce the voltage level by a voltage divider network, or second, to couple a tachogenerator (usually ac) to the output shaft of the motor. The output of the ac tachogenerator is rectified and attenuated to provide a suitable dc voltage representing the motor speed. In either case, this negative feedback voltage is compared against the voltage level at the output of the ramp generator. If the motor speed is lower than the desired speed, then the error voltage will cause the gate pulse generator (if a cosine-crossing method of firing control is assumed) to advance the firing delay angle and as a result to increase the motor speed. Because of the inertia of the drive and its connected load, the motor will accelerate above the desired speed. Once again, the error signal produced by comparing the desired and actual speed voltage signals will cause the firing delay angle to be retarded, and the drive will slow down. As can be seen, the motor will never remain constantly at the set speed, but will vary slightly about the desired speeu, ᴜᴇ amount of the variation is a measure of the speed regulating capability of the drive control system. Armature voltage speed sensing usually produces a speed regulating capability of ±2 percent, whereas an ac or dc tachogenerator speed sensor will provide a speed regulating capability of ±0.1 percent, which is more than adequate for most integral horsepower (kW) drive applications. If improved speed regulation is required, then digital speed sensing techniques are used.

Typical speed-torque curves of a constant torque drive are shown in Fig. 9-6, for a number of individual speed settings. As can be seen, these curves are a family of parallel curves, for each speed setting. The minimum speed is usually determined by the ability of the motor to dissipate heat in the absence of a rapid flow of cooling air through it. The limitation of the maximum torque is obtained from the current limit circuitry, since it is not a characteristic of the motor. It should also be noted that the speed regulation, which is the ability of the drive motor to maintain its speed under conditions of varying load, is usually defined in terms of the base speed, and as a result the speed setting is reduced. The drop in speed between no-load and full-load remains constant, but the actual speed regulation has increased.

In constant torque applications the motor field excitation remains unchanged, and as a result the dc excitation voltage is usually obtained from an

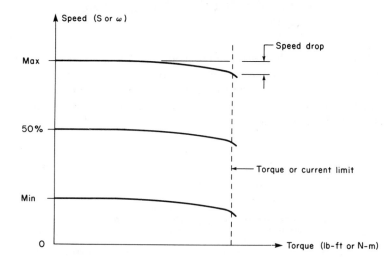

FIG. 9-6 Speed-torque curves of a typical dc adjustable speed drive.

uncontrolled rectifier connected across the ac supply lines, although in large horsepower (kW) machines, a separate controlled rectifier is preferred. This also ensures that under starting conditions maximum torque will be produced by providing full field excitation.

Most commercial SCR adjustable-speed drives are available in three-phase, full- or half-controlled configurations. They feature, usually as standard items, instantaneous electronic overload protection, phase loss and phase sequence protection, isolated control circuit power supplies, semiconductor and control circuit fuse protection, *R-C* snubber networks, etc. Optional controls usually are available, such as dynamic braking, inching, jogging, and plugging facilities, diagnostic metering, motor-overheating protection, field loss relay protection, process control followers, etc.

Although it has been common practice to use three-phase, half-controlled converters because of lower initial costs, the resulting worsened form factor of the mean output dc voltage has caused problems with motor commutation and overheating, with a consequent derating of the motor being required. If a three-phase full-controlled converter is used, nearly all dc motors can be driven without any derating or commutation problems.

Another obvious advantage of the three-phase, full-controlled converter is its ability to operate in the synchronous inversion mode and as a result to provide a regenerative braking capability in an overhauling load situation.

The full concept of the Ward-Leonard principle is utilized only in four-quadrant control schemes.

9-4   FOUR-QUADRANT CONTROL OR DUAL CONVERTERS

In dc motor control with a one-quadrant converter the power flow is from the ac source to the motor, and power cannot be returned from the motor to the ac source; that is, the converter cannot operate as a synchronous inverter. From Fig. 9-7, a one-quadrant converter can operate in quadrant 1, forward motoring, or in quadrant 3, reverse motoring.

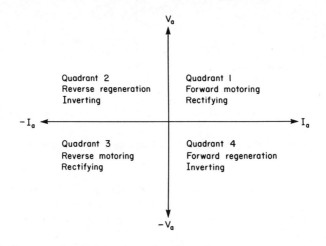

FIG. 9-7 Four-quadrant operation of a dc motor.

A two-quadrant converter has the capability of operating in the rectifying or inverting modes; that is, the motor is supplying the torque to the load (motoring), or the load is accelerating the armature in the same direction, and the motor is acting as a generator and is transferring power from the dc side to the ac side of the converter by synchronous inversion. This condition is known as regeneration or regenerative braking, and as a result the load accelerated armature will be slowed down. Under these conditions the converter can operate in quadrants 1 and 4 or quadrants 2 and 3.

There are many applications, such as a mine hoist or reversing steel mill drive, in which forward and reverse operation of the motor is required. There are several methods by which this may be accomplished: First, the output of a two-quadrant converter can be connected to the motor armature by means of a two-pole reversing contactor, which adds the complication of a time delay in the operation of the contactor, during which time control of the motor is lost. Second, the output of the converter to the dc motor remains unchanged, but the separately excited field is reversed, either by switching or by reversing the voltage applied to the field. The disadvantage of this method is that because of the time constant of the highly inductive field the response of the drive will be slow unless current in the field is forced by applying a voltage two or three times the field supply voltage. Third, the most effective method is to supply the motor armature circuit from two two-quadrant converters connected back-to-back, which achieves reversing control and regeneration without switching the armature circuit or reversing the field. This arrangement is shown in Fig. 9-8, where each of the converters is rated for the full load requirements of the motor.

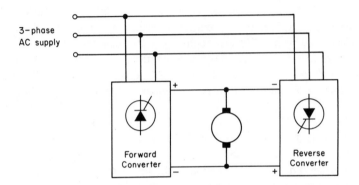

FIG. 9-8 A dual converter providing four-quadrant operation of a dc motor.

By simultaneous firing angle control of both converters very rapid reversals of speed and torque can be achieved. When converters are operated in this configuration, the combination is known as a *dual-converter*, and the current through the motor can flow in either direction. Ideally the firing delay angles of both converters are adjusted to give the same output voltage at their output terminals, and therefore

$$v_{d\text{POS}} + v_{d\text{NEG}} = 0 \tag{9-8}$$

where $v_{d\text{POS}}$ and $v_{d\text{NEG}}$ are the instantaneous voltages of the positive and negative converters, and

$$V_{d\text{POS}} + V_{d\text{NEG}} = 0 \tag{9-9}$$

where $V_{d\text{POS}}$ and $V_{d\text{NEG}}$ are the average voltages of the positive and negative converters.

Because the ac ripple components of the output dc voltages do not occur simultaneously, this condition defined by Eq. (9-9) cannot be met. As a result, in normal practice, even though the firing delay angles are varied simultaneously to meet the condition

$$\alpha_P + \alpha_N = 180° \qquad\qquad (9\text{-}10)$$

the control signals to the negative converter are blanked out when the motor armature current is positive, and those of the positive converter are blanked out when the motor armature current is negative. As a result, only one converter is in operation at a time, and the other presents a high impedance. The transition from motoring to regenerating, that is, from quadrant 1 to 4 or quadrant 2 to 3 or vice versa, takes place effectively in three steps. The current in the rectifying converter is reduced to zero by increasing the firing delay angle; then the gating signals are blanked out, and those of the other converter are reinstated and phased forward to increase the current from zero. The voltage at the armature remains unchanged in polarity. The converters are then operating in the "circulating current free" mode. A closed-loop control scheme for a dual converter is shown in Fig. 9-9. The actual speed signal derived from a tachogenerator coupled to the motor shaft is compared to the desired speed; the resulting error signal in turn is compared against a voltage signal representing the actual motor current. If the motor is not operating in current limit conditions, then the error signal is supplied to the steering logic. The steering logic determines which converter will be conducting, this being determined by an applied signal designating the desired direction of rotation of the motor armature, and second, by a current sensor in the armature circuit that senses the direction of current flow which will inhibit the changeover from the conducting converter until the current has become zero. Provided that all of these conditions are met, then a reference

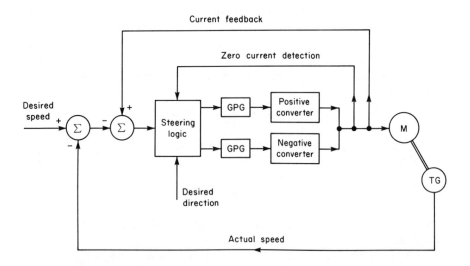

FIG. 9-9 Block diagram of a closed-loop control system for a dual converter dc motor drive.

signal will be applied to the appropriate gate pulse generator, which in turn will apply correctly sequenced firing pulses to the converter SCRs with the correct firing delay angle.

9-5   BRAKING

The only way to completely stop the rotation of a motor and hold its shaft in one position is by mechanical braking; however, the kinetic energy of the motor is dissipated as heat at the brake. There are three basic methods of slowing down a motor; these are plugging, dynamic braking, and regeneration.

9-5-1   Plugging

In plugging the voltage applied to the armature of a dc motor is reversed, or in the case of a three-phase ac motor, two of the phase connections are reversed. In both cases a counter-torque is produced which rapidly slows down the motor; if this torque is not sensed by a zero speed detector, the motor will accelerate in the opposite direction. During the plugging period the counter-emf of the dc motor is added to the plugging voltage, which can lead to overheating of the armature because of the high armature current.

The dual converter can provide plugging, by arranging that the conducting converter be suppressed and the other converter be operated in the rectifying mode. That is, the firing delay angle must not be greater than 90 deg, or the converter will be operating as a synchronous inverter, and regenerative braking will take place. The degree of plugging is controlled by varying the firing delay angle, the greatest rate of plugging occurring when $\alpha = 0°$.

9-5-2   Dynamic Braking

In dynamic braking, the stored kinetic energy of the motor uses the generator action of the motor to rapidly decelerate the armature. The basic sequence of events that must occur are, first, the armature circuit is isolated from the converter, or the converter is phased back so that its output voltage is zero; second, the armature output is connected across the dynamic braking resistor. At the same time, since the armature counter-emf is opposite in polarity to the output voltage of the converter, the armature current is reversed and attempts to reverse the armature. As a result, the motor slows down and uses up the stored energy of the system, which is dissipated in the dynamic braking resistance. When the armature stops rotating, all the energy is used up and the motor cannot reverse. The dynamic braking resistance is shunted across the armature terminals, either by a dynamic braking contactor or by a thyristor (see Fig. 9-10).

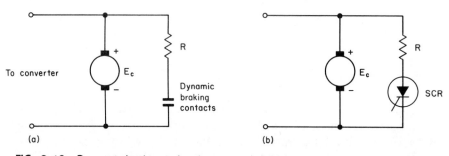

FIG. 9-10   Dynamic braking of a dc motor. (a) With a dynamic braking contactor; (b) with an SCR.

9-5-3   Regenerative Braking

Regenerative braking is similar to dynamic braking, but instead of dissipating the stored energy in a resistance, it is returned to the source. Under overhauling load conditions regeneration takes place. If the motor is being driven by the forward converter of Fig. 9-9, as the armature speed is increased, the counter-emf will increase, and the armature current will decrease to zero, and then start to increase in the opposite direction. When the armature current becomes zero, the forward converter is turned off, and the reverse converter is turned on with a firing delay angle greater than 90 deg, at which point it is operating as a synchronous inverter and the power transfer is from the dc motor armature to the ac source. The amount of power transferred is increased by increasing the firing delay angle toward 180 deg.

Regenerative braking is the most effective and energy-efficient system of braking, but requires more sophisticated control than dynamic braking. In both regenerative and dynamic braking, the field is maintained at full strength, and reliance is placed on the generated output of the motor for the braking action. As a result, the braking action is greatest when first applied and rapidly diminishes as the speed drops. For quick stopping, plugging is still the most effective. A combination of regeneration and plugging can easily be obtained in a dual converter, by initially permitting the power to be transferred to the ac supply by synchronous inversion; then advancing the firing delay angle to less than 90 deg will initiate plugging.

9-6   DC-DC DRIVES

The most energy-efficient method of controlling series dc motors operating from a dc source, e.g., a battery bank, or an overhead conductor, or a third rail system as used by electric trains, is the dc-dc or chopper control. Common applications of this principle are battery-operated delivery vehicles, golf carts, electric fork-lift trucks, street cars, trolley buses, and electric trains.

Control is achieved by varying the armature voltage and current. The tractive force is proportional to the motor torque, which in turn is proportional to the square of the armature current. Under steady-state load conditions the motor speed is inversely proportional to the field pole flux, which in turn in a series motor is proportional to the armature current. Typical series motor characteristics are shown in Fig. 9-11.

Prior to the introduction of chopper control, since the developed torque is at its greatest value at low speeds, starting acceleration was maximized by using series resistances in the armature circuit, which unfortunately produced a stepped acceleration; because of the energy dissipated in the controlling resistances, the motor operated with a low efficiency. Chopper control permits smooth control of the armature current and thus smooth acceleration without energy losses.

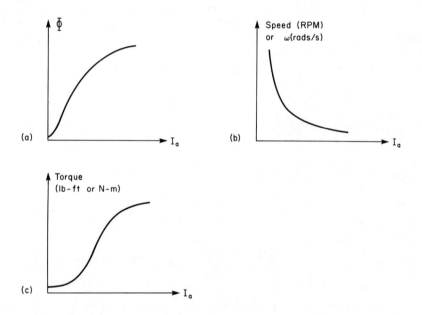

FIG. 9-11 Typical series motor characteristics.

9-6-1   The Jones Circuit

A basic problem in chopper control of a dc motor is the maximum armature current that can be commutated by the thyristor. As the motor horsepower (kW) increases, the worse case armature current will occur under locked armature conditions. This current can be calculated, or a locked armature test will determine the rate of change of the armature current. From this data the commutating circuit must be designed to achieve commutation before the main thyristor is subjected to the maximum armature current. (This problem is not significant in small motor applications.) To ensure commutation, a current feedback loop is usually used to sense the buildup of armature current. This circuit provides the limit control, without which the main thyristor would not commutate, and control would be lost.

In the Jones circuit the mean load voltage V_{do} is controlled by varying t_{ON} and t_{OFF}, i.e., pulse width modulation. The circuit as applied to a dc series motor is shown in Fig. 9-12(a), and its operation is as previously described in Sec. 7-2-3. The circuit can be modified to provide forward and reverse operation of a battery-operated vehicle when modified as shown in Fig. 9-12(b). In this circuit the forward-reverse direction switch is set for the desired direction. When the OFF-RUN switch is closed, if forward motion is desired, the forward relay FR is energized, and the normally open F contacts in the series field circuit close and the normally closed F contacts open. When SCR1 is turned on, current will flow through the field circuit from left to right through the normally closed R contact to the negative terminal of the battery. To reverse the vehicle it should be decelerated by reducing the on-time of SCR1, and then the direction should

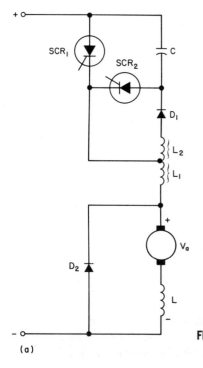

(a)

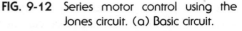

FIG. 9-12 Series motor control using the Jones circuit. (a) Basic circuit.

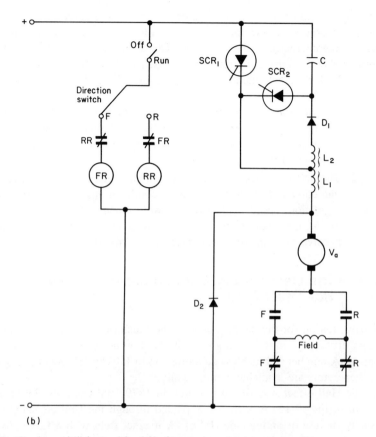

FIG. 9-12 (*cont.*) (b) modified for forward and reverse operation.

be changed to reverse. The forward relay *FR* drops out, and the reverse relay *RR* picks up; the contacts in the field circuit will cause the current to flow through the field from right to left, when SCR1 is turned on.

This section has discussed only the basic principle of two-direction control. In actual practice the vehicle would be equipped with acceleration control, braking, current limit, battery-saving circuitry, and safety interlocks to prevent a change of direction before the motor is stopped.

9-7   DC BRUSHLESS MOTORS

In spite of the variable-speed capabilities and the excellent speed-torque characteristics of conventional dc motors, the commutator limits the speed and voltage of dc machines, as well as being a continual maintenance problem. In recent years a good deal of research has been devoted to the elimination of the commutator and brushes by electronic means, while at the same time retaining the desirable characteristics of the dc motor. The result is the brushless dc motor. Unfortunately, this term is currently being used to describe two different types of motors with different characteristics and applications. The two types are:

1. The electronically commutated motor, which replaces the commutator and brush gear by solid-state switches, and retains the characteristics of the dc motor. It consists of a permanent magnet rotor with a wound stator and rotor position sensor.
2. The dc link converter and a synchronous-reluctance, synchronous or induction motor. Although these combinations provide good speed regulation, speed control over a wide range is less efficient. In addition, the rectifier-inverter and motor combination is bulky and expensive as compared to the brushless dc motor.

9-7-1   The Electronically Commutated DC Motor— Hall-Effect Type

The major reason for the development of the brushless dc motor was the need for high efficiency, long life, high reliability, low noise, and low power consumption. A number of brushless dc motors up to 1/40 hp (19 W) utilizing Hall-effect generators are available commercially.

The Hall effect was first discovered in 1879, and basically the principle is shown in Fig. 9-13. A current I is passed through the Hall element, which is usually indium or antimonide (InSb). A magnet field with a flux density B is applied at right angles to the element and causes the charge carriers to be redistributed within the element, causing a voltage V_H, the Hall voltage, to be

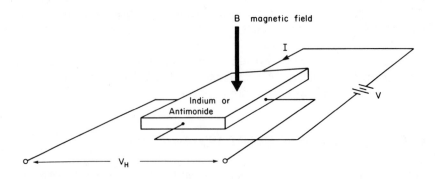

FIG. 9-13 Basic concept of a Hall-effect generator.

induced in a direction mutually perpendicular to the current and magnetic field. The Hall voltage is proportional to I and B and is

$$V_H = RIB/d \qquad (9\text{-}11)$$

where V_H = Hall voltage, volts
    I = current, amps
    B = flux density, Wbs/m^2
    d = element thickness, meters
    R = Hall constant

The application of the Hall generator is shown in Fig. 9-14. A permanent magnet rotor turns inside a four-pole stator in which are placed the stator coils $W1$ through $W4$. $A1$ and $A2$ are dc amplifiers that amplify the Hall voltage produced by the Hall generators $HG1$ and $HG2$. If the permanent magnet rotor is in the position shown, a Hall voltage is produced by $HG1$, which in turn is amplified by $A1$, producing a current flow through the stator winding $W3$. The resulting magnetic field causes the permanent magnet rotor to turn 90 deg clockwise. In turn, a Hall voltage is produced by $HG2$, amplified by $A2$ and then applied to the stator winding $W2$, and the rotor turns through another 90 deg to complete 180 deg of rotation. The magnetic field sensed by $HG1$ has reversed polarity, and a reversed-polarity Hall voltage is amplified by $A1$ and applied to stator winding $W4$. The rotor turns through a further 90 deg and lines up with $HG2$. Its voltage is amplified by $A2$ and applied to $W1$, causing the rotor to complete one complete revolution. As a result, a rotating magnetic field has been produced by the stator coils in the sequence $W3$, $W2$, $W4$, and $W1$.

In Fig. 9-14 the Hall generators were supplied by a constant current source, and as a result the rotor would turn at a constant speed. In order to achieve speed control the back emf of the motor can be sensed and compared to a reference voltage representing the desired speed. The difference voltage can be used to control the current through the Hall generator, and as a result speed control is obtained. It should be noted that the Hall voltage will be sinusoidal, since the rotor flux produces a sinusoidal distribution in the air gap. As a result, the transistorized amplifiers $A1$ and $A2$ operate in the amplification mode and produce a constant stator flux, which, combined with the constant rotor flux, produces a constant torque. Unfortunately, the losses in the transistorized amplifiers, since they are operating as amplifiers, will result in a relatively low overall efficiency, approximately 30 percent.

9-7-2   Photoelectronic-type DC Brushless Motor

An alternative method of electronic commutation is by the use of photodetectors to sense the rotor position. The basic concept is illustrated in Fig. 9-15, which shows one set of windings of a six-pole stator.

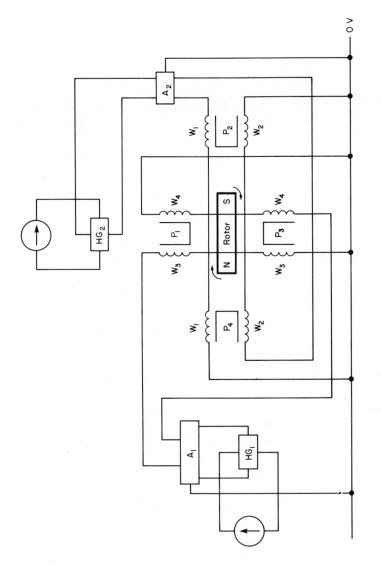

FIG. 9-14 Schematic of a four-segment Hall generator motor.

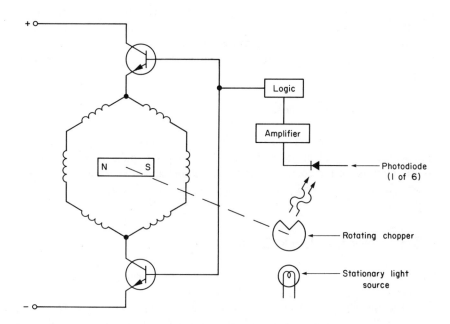

FIG. 9-15 Schematic of one segment of a photoelectrically commutated dc brushless motor.

The beam of a stationary light source is interrupted by a circular disc chopper mounted on the rotor shaft. This disc chopper has a 60-deg sector removed, and behind the chopper are mounted six equally spaced photodiodes. The photodiode outputs are amplified and in turn, via a suitable logic control, are applied to the bases of the two transistors, which are operating in the switching mode. As a result, current is applied in sequence to each of the six stator windings, producing a constant amplitude rotating field. The photodetectors, light source, amplifiers, logic, and switching transistors are mounted integrally with the motor. Since the transistors are operating in the switching mode, their losses are greatly reduced as compared to the Hall-effect motor, and the overall efficiency is much greater, of the order of 50 percent. The switching circuits can be modified to provide current limit, bidirectional control, and regenerative braking.

9-7-3 Summary

The major application of brushless dc motors is currently in the field of servomotors, magnetic tape, and disc drives, where they are used as direct drive motors, because of their high starting torque and precise speed control capabilities. The major disadvantage at present is their cost, which should drop considerably as volume production of Hall-effect sensors is achieved. The permanent magnet rotor under high starting current conditions may possibly be demagnetized; however, this effect is minimized by the use of high-impedance stator windings.

9-8   UNIVERSAL MOTORS

The series universal motor is extensively used in portable tools, in household appliances, such as blenders, and in small high-speed applications. Solid-state control circuits for the control of universal motors are simpler than those associated with dc motors and polyphase motors. In general, they are classified as half-wave or full-wave controls with or without feedback.

9-8-1   Nonfeedback Circuits

Since the series motor produces positive torque for either half-cycle of the ac supply, it is capable of operation under half- or full-wave control.

The cheapest and simplest half-wave control is shown in Fig. 9-16, which utilizes an R-C phase-shift circuit to vary the firing angle and a neon bulb as the trigger device to supply the gate pulse signal to the SCR. The major disadvan-

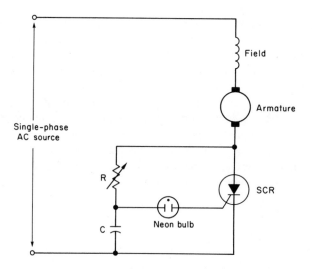

FIG. 9-16 Half-wave, nonfeedback, universal motor control.

tages of this circuit are, first, that there is only one torque pulse per cycle, and second, that the speed regulation is very poor.

The torque output of the universal motor can be greatly improved by using a TRIAC instead of the SCR in Fig. 9-16 to obtain full-wave phase control. A typical circuit is shown in Fig. 9-17. As before, phase-shift control is obtained by a simple *R-C* phase-shift network, using a DIAC as the trigger device. Increasing the value of *R* will increase the firing delay angle, and the motor will slow down.

The major problem with both of these circuits is that there is a rapid drop in speed as the load on the motor increases. The decrease in speed can be reduced by using a feedback circuit.

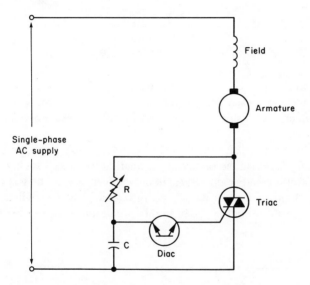

FIG. 9-17 Full-wave, nonfeedback, universal motor control.

9-8-2   Feedback Circuits

The simplest form of feedback circuit is the half-wave Momberg circuit shown in Fig. 9-18. A voltage signal V_d representing the desired speed is obtained by varying the speed-regulating potentiometer $R2$. The voltage V_a senses the armature counter-emf which is proportional to the motor speed. The diode D will conduct only when $V_d > V_a$, if V_d has been established for the desired speed. As long as the motor is running at the desired speed, D will block the gate signal to the SCR. If the motor speed drops, the armature feedback signal V_a decreases, and since $V_d > V_a$ a gate signal will be applied to the SCR. The

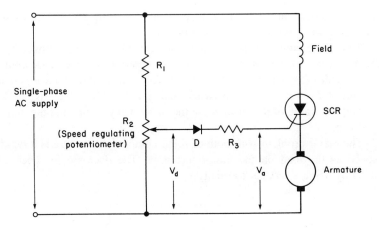

FIG. 9-18 The Momberg half-wave feedback circuit.

greater the drop in the motor speed, the earlier D will conduct in each positive half-cycle, and the greater will be the dc mean voltage applied to the motor armature, which will result in an increase in motor speed toward the desired speed.

The Momberg circuit has two major disadvantages: First, the final speed will be slightly less than the set speed. Second, at low-speed settings the range of speed control is greatly reduced. An improved feedback circuit that overcomes some of the disadvantages of the Momberg circuit is the Gutzwiller half-wave feedback circuit shown in Fig. 9-19. In this circuit the voltage across the speed-

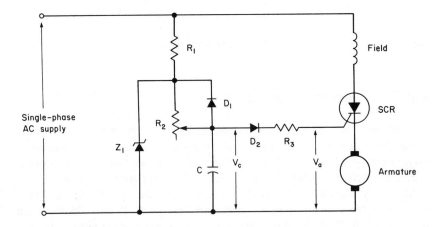

FIG. 9-19 The Gutzwiller half-wave feedback circuit.

regulating potentiometer $R2$ is established by the zener diode $Z1$. The voltage across the capacitor V_c is determined by the setting of $R2$; the blocking diode $D2$ will apply a gate pulse to the SCR whenever $V_c > V_a$. The greater the difference between V_c and V_a, the sooner in the positive half-cycle will the gate signal be applied to the SCR, and as a result the greater will be the mean dc voltage applied to the armature and the greater the speed.

The Gutzwiller circuit improves speed regulation, the range of speed control, and the stability of the motor, but at an increased cost of the control circuit components.

9-9   PHASE-LOCKED LOOP SPEED CONTROL SYSTEM

Until recently the major use of phase-locked loops was in the frequency synchronization of FM receivers, but now a great deal of attention is being devoted to their use in precise motor speed control.

Until the introduction of low-cost digital integrated circuits, the conventional method of speed control used analog devices such as the tachogenerator to provide feedback information. The advent of the integrated circuit using phase-locked loops permits control of speed within ± 0.002 percent of desired speed, with the added advantage that the concept can be applied to the control of motors of any size.

The basic concept of a phase-locked loop speed control system is shown in block diagram form in Fig. 9-20. It consists of a phase comparator or phase detector which compares a train of pulses generated by an encoder attached to the motor shaft with a train of pulses representing the desired motor speed generated by the voltage-controlled oscillator (VCO), which forms part of the phase-locked loop (PLL) chip. The PLL contains the comparator, a low-pass filter, and the VCO. The error signal produced by the comparator is filtered by the low-pass filter, whose function is to remove noise and control the dynamic characteristics of the drive, producing an output signal that is used to vary the VCO frequency and supply a signal to the dc power amplifier.

The VCO is basically an astable multivibrator whose frequency is set by an external R-C combination to the center or free-running frequency. The operation of the system depends upon the stability of the VCO center frequency. The feedback signal from the encoder representing actual speed is locked to the VCO frequency, provided that it is within the capture range. Any drift of the VCO frequency caused by aging, noise, or temperature variations will be followed by the motor. If there is a deviation of motor speed, this deviation will be detected by the comparator, and a dc error signal will be produced by the low-pass filter, which will cause the VCO frequency to stay in phase with the encoder signal, and will also apply a corrective signal to the dc amplifier, whose output will be varied to oppose the change in the motor speed. As a result, the

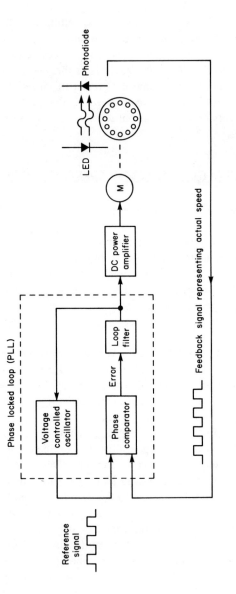

FIG. 9-20 Block diagram of a phase-locked loop applied to dc motor speed control.

motor is forced to maintain the speed set by the center frequency of the VCO. The desired speed is set by varying the center frequency of the VCO.

It is common practice in phase-locked loops to lock the VCO center frequency at the mean desired operating frequency, and then add an external bias voltage to the varying dc error voltage. This bias then can act as the setpoint to provide the desired frequency variation. As long as the frequency excursions do not exceed the lock range of the PLL, the VCO will follow.

The major advantages of phase-locked loop techniques are precise speed control and the extremely low cost of the control components, as compared to a conventional analog speed control system.

9-10   DC MOTOR SPEED CONTROL BY A MICROPROCESSOR

An alternative approach to digital speed control of a dc motor is by the use of a microprocessor. The basic technique is similar to that used in the phase-locked loop technique, except that the motor, encoder, and motor driver (in our example a chopper) replace the VCO and low-pass filter. In constant-speed applications it is desirable for the motor to have a high rotor inertia, since the rotor inertia will minimize speed variations because of the flywheel effect.

The basic block diagram of such a system is shown in Fig. 9-21. The basic operation of the system is as follows: The actual motor speed is obtained by measuring over a discrete interval of time, e.g., 200 msec, the number of pulses produced by the encoder coupled to the output shaft of the motor. This signal representing the actual speed is compared with the set speed entered into the microprocessor. The output of the comparator is the error signal, which in turn is applied to the gate pulse generator. The input to the gate pulse generator consists of two pulse trains, one of which controls the gating signal to the main SCR in the chopper and initiates the t_{ON} pulse to the motor. The other pulse train, which controls the commutating SCR, is phase-shifted with respect to the first by an amount proportional to the error between the desired and actual speeds. In other words, if the motor is running slow, then the phase shift between the two pulse trains will be increased so that the t_{ON} period of the chopper will be increased; the mean dc output voltage will also be increased, and the motor will speed up. The chopper will be operated so that t_{ON} and t_{OFF} are variable, but the periodic time is constant; that is, the chopper is being operated in the pulse width modulation mode.

Unlike the phase-locked loop system, the microprocessor only samples the actual speed at discrete intervals. The desired speed is usually entered in as a 12-bit binary number by means of toggle switches, a thumbwheel switch, or a computer keyboard terminal. The counter is a software program which produces a 12-bit binary number representing actual speed, which is compared

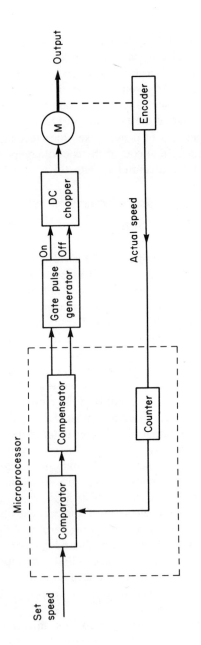

FIG. 9-21 Block diagram of a microprocessor speed control of dc chopper-fed motor.

against the 12-bit binary number representing the desired speed. The resulting error signal determines the magnitude of the phase shift between the two pulse trains applied to the chopper SCRs. This system can be scaled up to control large-size dc motors.

9-11   AC MOTOR CONTROL

The major types of ac motors that are normally associated with thyristor control are polyphase induction motors [both squirrel cage (SCIM) and wound-rotor (WRIM)], and synchronous-reluctance and synchronous motors. All of these polyphase machines can be speed-controlled by using either the dc-link converter or the cycloconverter; however, in addition the wound-rotor induction motor can also be speed controlled by control in the rotor circuit.

Solid-state control techniques are being applied to motor control in applications such as soft starting, motor protection, controlled acceleration for pumps and conveyors, reduced voltage starting, constant torque acceleration, linear acceleration and deceleration, as well as jogging, reversing, controlled plugging, controlled plug stop, and dynamic braking.

Solid-state static inverters and converters control ac motor speeds by varying either the frequency or the voltage applied to the stator, or both, through conversion of ac to dc and back to ac at the desired frequency by the dc-link converter, or, in the case of the phase-controlled cycloconverter, by direct ac-to-ac frequency conversion.

9-11-1 Variable-frequency Speed Control

The most commonly used ac drive system is the variable-frequency dc-link converter and polyphase induction motor or polyphase synchronous motor. The operating characteristics of either of these motors are retained over the frequency range of the inverter, typically from 10 Hz to 200 Hz. Major applications of variable-frequency drives are textile machinery, machine tools, steel and paper mill equipment, and electric traction.

When a balanced three-phase ac supply is applied to the stator of a three-phase machine, a constant amplitude rotating magnetic field is produced. The angular velocity of this field is given by

$$S = \frac{120f}{P} \text{ rpm} \qquad (9\text{-}12\text{E})$$

or
$$\omega = \frac{4\pi f}{P} \text{ rad/sec} \qquad (9\text{-}12\text{SI})$$

where S or ω = synchronous speed of the rotating magnetic field
$\qquad$ f = frequency, Hz
$\qquad$ P = number of stator poles/phase

$\quad$ In turn it can be seen that the synchronous speed of the rotating magnetic field is proportional to the frequency of the supply.

$\quad$ In the case of the polyphase induction motor, the actual rotor speed is slightly less than the synchronous speed of the rotating magnetic field. Since the rotor torque is caused by the interaction of the magnetic field produced by the induced rotor currents and the rotating magnetic field, the speed difference is called the *slip*, and is

$$s = \frac{S - S_R}{S} \tag{9-13E}$$

or $$s = \frac{\omega - \omega_R}{\omega} \tag{9-13SI}$$

As previously developed in Chapter 6, Eq. (6-14), the internal torque is

$$T = \frac{mP}{4\pi} \cdot \left[\frac{E_1}{f}\right]^2 \cdot \frac{f_r r_2}{[r_2^2 + s^2 x_2^2]}$$

from which it can be seen that the internal torque developed is proportional to E_1/f. The air-gap flux must be constant at all frequencies in order to produce a constant torque output. As a result, the ratio E_1/f must also be constant, in order to produce a constant air-gap flux. If stator leakage impedance is small, then E_1 is proportional to the applied stator voltage V_1, and the air-gap flux will also be constant when V_1/f is constant.

$\quad$ At low frequencies, i.e., approximately 10 Hz, the phase resistance is the major component of the phase impedance, and as a result there will be a decrease in the air-gap flux at low frequencies. If this condition is not acceptable, the constant voltage/Hz ratio must be modified to restore the air-gap flux, even though there is a possibility that the stator iron is in saturation.

$\quad$ The breakdown or maximum torque is constant for a given machine at all frequencies and is proportional to the air-gap flux squared and inversely proportional to the rotor leakage reactance. The only effect of the rotor resistance is to determine the slip at which the breakdown torque is developed. Under constant air-gap flux conditions the starting torque is greater at all frequencies than when supplied at rated frequency and voltage (see Fig. 9-22).

$\quad$ In summary, operating a polyphase induction motor, either a squirrel cage or wound rotor, with a constant volts/Hz ratio, will result in higher starting and

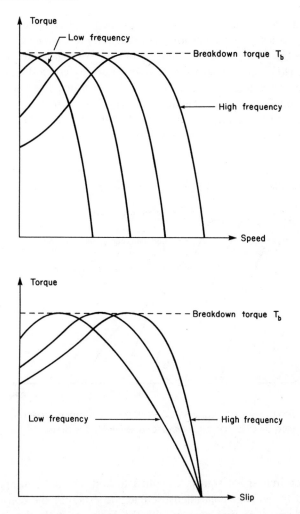

FIG. 9-22   Torque-speed and torque-slip curves of a polyphase induction motor with a variable frequency supply.

breakdown torques, and with the same full load slip the torque is greater at the higher frequencies. Additionally, the horsepower (kW) output and efficiency are greater at the higher frequencies. Failure to maintain the constant volts/Hz ratio will affect the constant torque output by the square of the air-gap flux density, or will cause the stator current to increase and overheat the motor.

In most variable-frequency drives the constant volts/Hz ratio is maintained up to the rated frequency of the motor and then the stator voltage is maintained at its rated value as the frequency is increased. As a result, the motor operates in a constant torque mode up to rated frequency, usually 60 Hz, and then operates in a constant horsepower (or constant kW) mode above rated frequency (see Fig. 9-23). Modern dc-link converters can be programmed to provide the constant volts/Hz ratio for constant torque control by controlling the amplitude of the output ac voltage, using pulse width control or pulse width modulation control, although pulse width modulation control is the preferred method, since it minimizes the harmonic content of the output ac voltage. The basic control requirements of a variable-frequency inverter are shown in schematic form in Fig. 9-24.

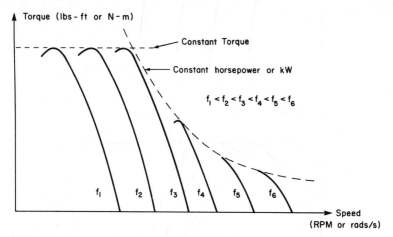

FIG. 9-23   Torque-speed curves for variable frequency operation in constant torque and constant horsepower (kW) modes.

The desired speed signal is applied to both the pulse width modulation control and the voltage-controlled oscillator to establish the constant volts/Hz ratio; the three-stage ring counter, via the logic, controls the distribution of gating pulses to the six-step inverter. During each half-cycle that the thyristors are gated on, the pulse width modulation control will ensure that the thyristors are switched on and off to achieve voltage control within the inverter. A voltage feedback permits control in the constant torque or constant horsepower (kW) modes, while at the same time control of the stator current is obtained by means of the current feedback to the pulse width modulation control. Because of the different dc potentials at the gates of the thyristors, it is essential that isolation be provided, either in the form of pulse transformers or optocouplers.

Under inductive loading conditions the load current is not immediately transferred to the incoming thyristors after commutation, but is flowing through the feedback diodes. However, with PWM the incoming thyristors will be gated

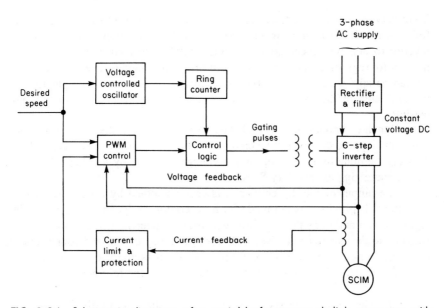

FIG. 9-24 Schematic diagram of a variable-frequency dc-link converter with pulse width modulation.

on when the current reverses. Under low-frequency operating conditions the cores of the isolating pulse transformers would have to be large, a problem which can be overcome by using optocouplers.

Because of the essentially constant speed characteristics of polyphase induction motors, unless very precise speed control is required, it is normal practice to use open-loop control techniques and rely on the stability of the voltage-controlled oscillator.

If constant speed is essential, then a speed signal must be brought back to modify the control oscillator output. Alternatively, a constant speed output can be obtained by using either a synchronous-reluctance or synchronous motor whose rotors rotate in synchronism with the rotating magnetic field.

Under 5 kVA it is standard practice to use a transistorized inverter, which simplifies the control by eliminating the commutation control circuitry. It is also expected with the recent introduction of improved power transistors that the kVA capacity of transistor inverters will be significantly increased. For high-kVA outputs up to 200 kVA, thyristor inverters using fast-switching, inverter-grade thyristors are used. If regenerative braking is desired, the dc rectifier must be a full-controlled rectifier in order that power may be returned to the ac source, and since the current flow is reversed when the inverter is returning power, which will occur when the frequency is rapidly lowered or under overhauling load conditions, it will be necessary to reverse the rectifier output leads. This is not very satisfactory, and, as a result, it is normal practice to connect a second full-controlled converter in inverse parallel with the rectifier and use it as an inverter to achieve regeneration (see Fig. 9-25). The converter acting as the inverter for regeneration is prevented from firing until there is reversal of current. As can be appreciated, this configuration increases the control complexity and the total cost of the installation, and is only used if dynamic braking is not feasible.

Dynamic braking is accomplished by connecting a discharge resistance across the input to the inverter when the dc voltage rises because of regeneration. This can be done by the use of an overvoltage relay, or by a voltage detection circuit energizing an auxiliary SCR.

The output of the dc rectifier is always smoothed by a capacitor across the output. If provision is not made for regeneration or dynamic braking, under regenerative conditions the capacitor will become overcharged and the voltage supplied to the inverter will rise, with a resulting rise in iron and copper losses of the machine, which will indirectly provide a substitute for dynamic braking.

The major advantages of dc-link converters are the following:

1. The steady-state speed accuracy is excellent, typically 0.05 percent over a 10:1 speed range.
2. The dynamic response is very rapid.
3. The power factor at full load is usually about 0.95.
4. The efficiency is 85 percent or better.
5. Installation costs are relatively low.

Reduced voltage starting is unnecessary, since the frequency is reduced to its minimum value and then increased to its desired value, thus achieving motor starting without high inrush starting currents. In addition, multimotor control is easily obtained from one inverter, permitting synchronized control of a conveyor line, or, in the case of electric traction, control of a number of traction drive motors. Where precise speed control is required, the synchronous or synchronous-reluctance motor is used.

The other major method of static frequency conversion is the cycloconverter, which is an ac-to-ac frequency converter; it inherently operates at a higher efficiency than the dc-link converter. It has two major disadvantages: First, in order to maintain the harmonics in the output waveform at an acceptable level, it must be operated in the region from 0 Hz to one-third of the ac source frequency. Second, for bidirectional operation a three-phase cycloconverter requires a minimum of 36 SCRs with the attendant complexity in the control circuitry.

The cycloconverter is both frequency- and voltage-controlled by a cyclic variation of the firing delay angles, the output voltage being a replica of the

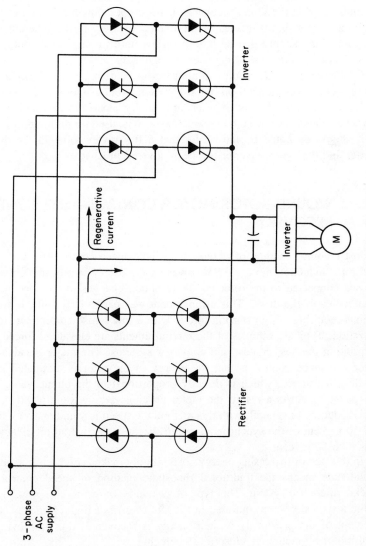

FIG. 9-25 Dual converter configuration for regeneration of a variable frequency ac drive.

3 - phase AC supply

Rectifier

Inverter

Regenerative current

Inverter

M

reference voltage. Reversal is easily obtained by reversing the polarity of the reference signal. Regeneration is an inherent property of the cycloconverter, and this, combined with the ability to reverse, permits operation in all four quadrants.

Some of the disadvantages of the cycloconverter are the inherently low power factor and supply line pollution; furthermore the maximum output frequency is usually one-third of the source frequency.

Cycloconverters are finding application in gearless drives. Currently the largest system drives an 8500-hp (6400-kW) synchronous motor at 14 rpm used to drive a cement kiln. Cycloconverters are equally capable of driving induction and synchronous-reluctance motors, and may also be used for multimotor drives.

It should be noted by the reader that a full review of both the dc-link converter and the cycloconverter has been covered in Chapter 6.

9-12   WOUND-ROTOR MOTOR CONTROL—SLIP POWER RECOVERY

The most common method of obtaining variable-speed control of a polyphase wound-rotor induction motor (WRIM) was by the use of an adjustable external resistance connected to the rotor circuit to obtain changes in the speed-torque characteristics of the drive. This arrangement was particularly suitable for applications requiring a high starting torque with a low inrush current, or smooth acceleration. Typical examples of these requirements are elevators, forced draft fans, printing presses, cranes, and conveyor systems. The major disadvantage of rotor resistance control is that the energy in the rotor is dissipated in the controlling resistance, which is directly proportional to the ohmic value for a given load torque. As a result, the motor efficiency drops significantly as the speed is reduced. In general, with this method the speed is usually never reduced below 50 percent of the synchronous speed, because the overall efficiency will be less than 50 percent.

In this age of increasing energy costs, it only makes good sense to convert by solid-state means the traditional rheostatic method of speed control to a regenerative control system. This type of system is known as slip power recovery, and a typical system, outlined in Fig. 9-26, gives a high efficiency. Up to 98 percent of the previously wasted energy is recovered, thus improving the overall motor efficiency by close to 25 percent.

The basic principle of operation depends upon the rotor emf being reduced and as a result the rotor current is reduced. The reduction in rotor current will cause a decrease in the rotor speed in order that the induced rotor emf will increase to produce an increased rotor current to meet the torque requirement at the lower speed. Effectively the inverter injects an emf into the rotor circuit that is anti-phase to the rotor emf.

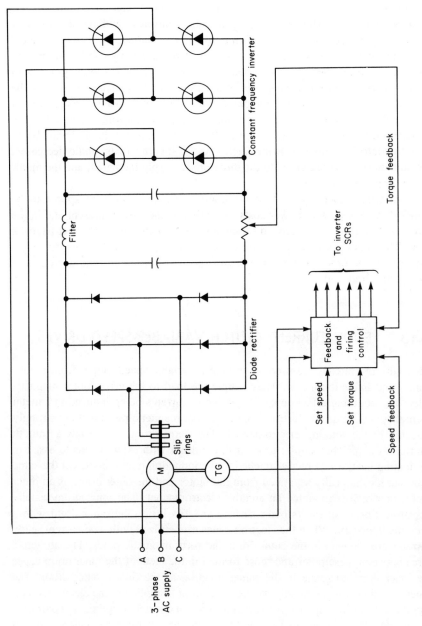

FIG. 9-26 Slip power recovery speed control of a wound rotor induction motor.

The operation of the slip power recovery system depicted in Fig. 9-26 is as follows: The ac rotor emf is rectified by a diode bridge and supplied to a three-phase inverter whose SCRs are gated in synchronism with the stator supply voltage. By varying the inverter firing angle, the amount of power extracted from the rotor circuit and returned to the three-phase supply can be controlled, and thus the rotor speed can be varied. To operate as a closed-loop system with variable speed and torque capabilities requires that input signals representing the desired speed and torque be compared with the actual speed derived from a tachogenerator coupled to the motor shaft, and the actual torque derived as the voltage drop across a resistor in the dc-link circuit, the resulting signals being used to control the inverter firing angle. As the inverter firing angle decreases, the amount of power returned to the three-phase supply increases, and the motor speed drops, or vice versa.

At first glance the cost of the control system would not appear to be justified; however, with power costs escalating as they are at present and in the foreseeable future, the improved efficiency results in a relatively short payback time. The major disadvantage of the system is its inherently poor power factor, which is of the order of 0.5 at maximum speed and full load, and decreases as the speed is reduced.

9-13   EDDY CURRENT CLUTCH VARIABLE-SPEED DRIVES

Eddy current clutches are used to provide a variable-speed output from a constant-speed induction motor. They couple the load to the drive motor magnetically. The coupling effect depends upon eddy currents being induced in a metal drum that surrounds an electromagnet, creating magnetic poles that are strongly attracted to the rotating electromagnet. The outer drum, which has a smooth interior, is somewhat similar to a car brake drum with external fins to assist in dissipating heat; it is directly attached to the induction motor shaft. On the same axis, but mechanically separated from the motor and the outer drum, is an inner drum, in which is mounted an annular electromagnet. The inner drum usually has projecting teeth on its outer surface and fits inside the outer drum with a very small air-gap. When the electromagnet is energized with a dc current, eddy currents are induced in the outer drum and form magnetic poles. The attractive force between these poles and those formed in the teeth of the inner drum cause the inner drum to rotate in the same direction as the driven outer drum. The inner coupling will rotate only when a dc current is applied to the electromagnet, and the amount of slippage between the inner and outer couplings is controlled by the dc excitation applied to the electromagnet. It must be understood that even with full excitation applied to the electromagnet there must be a slippage between the inner and outer couplings in order that the eddy currents may be induced. A typical closed-loop control of an eddy current clutch system is shown in Fig. 9-27.

The set speed voltage is compared with the actual speed voltage derived from a tachogenerator coupled to the load. The resulting error signal produced by the error amplifier is applied to a dc power amplifier, and its output in turn is applied directly to the electromagnet coil in the eddy current clutch.

Eddy current clutches provide a wide range of output speeds, permitting the drive motor to be started under no-load conditions, and can be operated under constant torque conditions. The input power to the electromagnet is relatively low, for example, as little as 250 W for a 50-hp (37-kW) drive. Eddy current clutch drives are available in all horsepower (kW) ranges from ¾ hp (0.6 kW) up to as high as 18,000 hp (13,500 kW) on special order. Typical applications of eddy current drives are blowers, compressors, conveyors, cranes, dredges, elevators, line shafts, winders, etc. They have the advantage that they are easy to maintain, are rugged, and provide an excellent substitute for the static inverter in the speed range below the base speed of the motor.

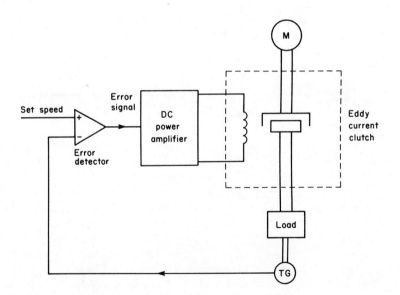

FIG. 9-27 Closed-loop feedback control of an eddy current clutch drive.

9-14   STATOR VOLTAGE CONTROL OF SPEED

The output speed of polyphase motors is determined by the supply frequency; in the case of the synchronous motor its speed is solely determined by the source frequency. The polyphase induction motor, on the other hand, when supplied at a constant voltage from a constant frequency source will run at a subsynchronous speed, usually of the order of 95 percent of the synchronous speed. The torque developed by the motor is proportional to the square of the applied voltage, and as a result under steady-state load conditions the motor torque adjusts itself to balance the torque demanded by the load. As the stator voltage is reduced, the rotor speed decreases; however, the power dissipated by the rotor circuit must increase in order to maintain a constant torque output. With a normal rotor and a constant torque load this method of speed control results in excessive losses and poor efficiency, but the losses may be reduced somewhat by the use of a high-resistance rotor, NEMA Class D. The system is most suitable in controlling fan and pump loads where the load torque is approximately proportional to the square of the speed. A typical system is shown in block diagram form in Fig. 9-28. The actual speed signal produced by the

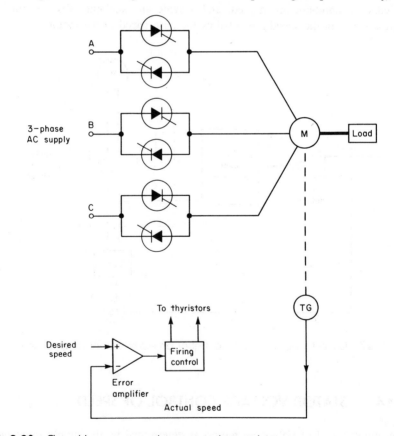

FIG. 9-28 Closed-loop stator voltage speed control system.

tachogenerator coupled to the motor is compared with the desired speed signal, and the output of the error amplifier is used to control the firing delay of the inverse-parallel-connected SCRs in series with the motor stator. If there is an increase in the load torque, the resulting speed decrease is detected by the tachogenerator. The resulting error signal will reduce the firing delay angle of the SCRs and increase the stator voltage and speed, and vice versa. This circuit can also be modified to give a reduced voltage starting capability by the intro- duction of a ramp generator, which will control the rate of acceleration of the motor to the desired speed. Other options can also be supplied, such as regen- eration, zero voltage switching, reversing, current limit, and protective features. The major disadvantage of reduced stator voltage speed control techniques is that they are very wasteful of energy. With the increasing cost of electricity it is certain that more attention will be given to the use of energy-efficient speed control methods, such as static inversion systems.

9-15   VARIABLE-FREQUENCY SYNCHRONOUS MOTOR DRIVES

In high-horsepower (-kW) applications the synchronous motor is very often used in combination with a variable-frequency static converter, e.g., with a cyclo- converter as a gearless drive for a cement kiln. This drive is a variable-fre- quency, 0–5-Hz, 8600-hp (6500-kW), 40-pole synchronous motor rotating at 14 rpm, using a cycloconverter supplied from a 60-Hz source.

In general, conventional synchronous motors are used in high-horsepower (kW) applications where the slip-rings do not present a hazard. Apart from maintaining a constant speed with a constant frequency source, they have the added advantage of being able to generate a leading or lagging power factor by proper adjustment of the dc rotor current.

In slow-speed operation the rotor will be large in diameter and short in axial length with a large number of salient poles, and will have a poor dynamic response. Higher-speed machines, on the other hand, require fewer poles; the rotor will be reduced in diameter and increased in axial length, and as a result the dynamic performance will be improved. Normally the stator is supplied from the static frequency converter, and the rotor will turn in synchronism with the rotating magnetic field. The dc-energized rotor is normally supplied by a phase- controlled rectifier via the slip-rings.

The synchronous motor may be operated in the motoring or regenerative mode, provided that there is a path for regenerative power to be returned to the ac source. For four-quadrant operation the synchronous motor must be brought to zero speed before the reversal of the stator connections takes place, or the machine will pull out of synchronism. This is accomplished by reducing the inverter frequency to zero and then building it up to the desired frequency again after reversing the stator connections.

The conventional synchronous motor drive is not self-starting, and it is normally started as a synchronous-induction motor and brought up to slip speed, prior to energizing the rotor field. With solid-state control the dc rotor field is energized at the commencement of the starting process, and the frequency of the stator supply is increased from 0 Hz up to the desired value; the rotor will stay in synchronism during the acceleration to rated speed.

9-16   SUMMARY

Solid-state drive systems, either ac or dc, have been making tremendous inroads into every area of the electrical and manufacturing industries. In the case of dc drives, the capability of phase-controlled converters to handle any size of drive, combined with improved reliability and reduced maintenance costs, has maintained the areas of precise speed control and rapid reversing capabilities for the dc motor in the steel and paper industries.

The tremendous advances that have been made in static frequency control, combined with the lower initial and subsequent maintenance costs of the polyphase induction motor, are resulting in an ever-increasing appearance of the variable-speed ac drive system in all fields where traditionally the dc drive was superior.

The continuing decline in the cost of thyristors, together with their increased power capabilities and high efficiencies, is seeing the application of the thyristor in most fields where electromagnetic control was the only method that was considered.

With the energy crisis becoming more apparent, it is expected that there will be a great increase in electric transit systems, and once again solid-state techniques are being used as the sole type of control.

Industrial processes are combining computer or microprocessor control in ever-widening fields of activities in the control of large systems.

REVIEW QUESTIONS

1. Discuss the requirements that must be met to achieve speed control of a dc motor above and below base speed—also include limitations and applications.
2. What is the function of a ramp generator in a closed-loop dc motor speed control system?
3. Describe the operation of a typical ramp generator as applied to static dc motor control.

4. What control features would you expect to find on a commercial three-phase static converter for a dc motor drive system?

5. What is meant by a four-quadrant or dual converter? Discuss the relative merits of the means by which this control may be achieved?

6. Detail the control requirements of a dual converter system and the merits of the "circulating current free" mode of operation?

7. Explain plugging, dynamic, and regenerative braking. How can these methods be incorporated into an SCR controller for a dc drive system?

8. What are the benefits of using a dc-dc drive control of a series dc motor?

9. Describe an application of the Jones circuit to a dc series motor control.

10. What are the advantages and disadvantages of dc brushless motors? Suggest typical applications.

11. Describe with the aid of a sketch the principle of operation of a dc brushless motor controlled by Hall-effect transducers.

12. Describe a typical photoelectric-type control of a dc brushless motor.

13. Discuss with the aid of a schematic the application of the Momberg circuit to the control of a universal motor.

14. Discuss with the aid of a schematic the application of the Gutzwiller circuit to the control of a universal motor.

15. What are the advantages of using a phase-locked loop speed control system?

16. With the aid of a block diagram describe the operation of a phase-locked loop speed control system as applied to a dc motor.

17. Describe an application of the microprocessor to the control of a dc motor drive.

18. What are the requirements that must be met by a dc-link converter when regenerative braking of the polyphase induction motor is desired?

19. What are the advantages in using a dc-link converter for the control of polyphase motors?

20. Describe with the aid of a schematic the speed control of a wound-rotor induction motor by slip power recovery? List the advantages and disadvantages.

21. Describe the principle of operation of an eddy current clutch.

22. Describe the principle of polyphase induction motor speed control by stator voltage control. List the advantages and disadvantages?

23. Analyze the application of polyphase cycloconverters in gearless drive high-horsepower (kW) synchronous motor applications?

Glossary of Symbols

A	Amperes
d	Element thickness, Hall generator, meters
E_c	Back or counter-emf, volts
E_x	Voltage drop caused by commutation overlap, volts
E_1	Counter-emf per phase produced by resultant airgap flux, volts
f	Frequency, hertz (Hz)
f_b	Rotor breakdown frequency, hertz
f_r	Frequency of induced rotor voltage, hertz
I	Current, amperes
I_a	Armature current, amperes
I_d	Phase-controlled converter output current, amps
I_L	Load current, amperes
I_m	Magnetizing current per phase, amps
I_0	No-load current per phase, amps
I_1	Stator current per phase, amperes
I_2	Rotor current per phase, amperes
K	Constant of proportionality $(ZP/60a) \times 10^{-8}$, English units
k	Prefix multiplier kilo or $\times 1000$

k	Constant of proportionality $(ZP/2\pi a)$, SI units
kV	Kilovolts
L	Self-inductance, henrys
l_2	Rotor leakage inductance per phase, henrys
M	Prefix multiplier mega or $\times 10^6$
MW	Megawatts
m	Prefix multiplier milli or $\times 10^{-3}$
mA	Milliampere
mW	Milliwatt
P	Number of stator poles/phase
P_{ROT}	Rotational losses, watts or kilowatts
R	Hall constant
	Resistance
R_a	Armature resistance, ohms
RCL	Rotor copper loss, watts or kilowatts
RPD	Rotor power developed, watts or kilowatts
RPI	Rotor power input, watts or kilowatts
RPO	Rotor power output (mechanical), watts or kilowatts
r_1	Effective stator resistance per phase, ohms
r_2	Rotor resistance per phase, ohms
S	Synchronous speed, revolutions/minute
	Angular velocity or speed, revolutions/minute
S_r	Rotor speed, revolutions/minute
s	Slip of rotor or speed below synchronous speed in decimal value or percent
T	Periodic time
	Temperature, degrees C (°C)
	Torque or turning effort, newton-meters
T_b	Breakdown torque, newton-meters
t	Torque ratio
	Time, seconds
t_{fr}	Circuit recovery time, μsecs
t_{ON}	Duration of chopper voltage pulse
	Integral number of cycles thyristor is conducting during pulse burst modulation
t_{OFF}	Off period of chopper voltage pulse

	Integral number of cycles thyristor is nonconducting during pulse burst modulation
V_a	Armature voltage, volts
V_c	Voltage drop across capacitor, volts
	Stored voltage on capacitor, volts
V_d	Average value of the input dc voltage to an inverter or chopper, volts
V_{do}	Average value of the output dc voltage from a chopper, or a converter when $\alpha = 0$ deg.
$V_{do\alpha}$	Average of the output dc voltage from a converter with a firing delay angle α
V_H	Hall voltage, volts
V_{OUT}	Output voltage, volts
V_R	Voltage drop across resistance, volts
	DC reference voltage, volts
V_{S1}, V_{S2}	Two-pulse midpoint converter transformer secondary voltages, volts
V_1	Applied stator emf per phase, volts
W	Power, watts
X_c	Commutating reactance, ohms
x_1	Stator leakage reactance per phase, ohms
x_2	Rotor standstill reactance per phase, ohms

GREEK SYMBOLS

Name	Capital	Lower Case	Use or Definition
Alpha		α	Converter firing delay angle, measured from the point at which the converter operates as if it were an uncontrolled rectifier
Beta		β	Inverter advance angle, the angle in advance of the point at which commutation is commenced. $\beta = 180° - \alpha$
Gamma		γ	Conduction angle, degrees
			Retardation angle, degrees
Delta	Δ		Variation or change of
			Change of temperature, degrees
Delta		δ	Output pulse width
Eta		η	Efficiency in decimal or percent
Mu		μ	Micro or 10^6
		μsec	Microseconds
		μ	Commutation overlap angle
Pi		π	Ratio of circumference to diameter, 3.141593
Sigma	Σ		Summation
Phi	Φ		Magnetic flux maxwells or lines
			Magnetic flux webers
Omega		ω	Angular velocity, radians/second
Omega		ω	Synchronous speed, radians/second

Bibliography

Bedford, B. D. and R. G. Hoft, *Principles of Inverter Circuits*. New York: John Wiley & Sons, Inc., 1964.

Daniels, A. R., *The Performance of Electrical Machines*. London: McGraw-Hill Publishing Co., Ltd., 1968.

Davis, R. M., *Power Diode and Thyristor Circuits*. Cambridge: Cambridge University Press, 1971.

Dewan, S. B., and A. Straughen, *Power Semiconductor Circuits*. New York: John Wiley and Sons, Inc., 1975.

Fink, Donald G. and H. Wayne Beaty, *Standard Handbook for Electric Engineers,* 11th ed. New York: McGraw-Hill Book Company, 1978.

Fitzgerald, A. E., Charles Kingsley, Jr., and Alexander Kusko, *Electric Machinery*, 3rd ed. New York: McGraw-Hill Book Company, 1971.

Gentry, F. E., F. W. Gutzwiller, Nick Holonyak, Jr., and E. E. Von Zastrow, *Semiconductor Controlled Rectifiers, Principles and Applications of p-n-p-n Devices*. Englewood Cliffs, N.J.: Prentice-Hall, Inc., 1964.

Kusko, A., *Solid-State D.C. Motor Drives*. Cambridge, Mass. and London: The M.I.T. Press, 1969.

Kosow, Irving L., *Control of Electric Machines*. Englewood Cliffs, N.J.: Prentice-Hall, Inc., 1973.

Mazda, F. F., *Thyristor Control*. London: Newnes-Butterworth, 1973.

Murphy, J. M. D., *Thyristor Control of A.C. Motors*. Oxford: Pergamon Press Ltd., 1973.

Pelly, B. R., *Thyristor Phase-Controlled Converters and Cycloconverters, Operation, Control and Performance*. New York: John Wiley and Sons, Inc., 1971.

Ramshaw, R. S., *Power Electronics, Thyristor Controlled Power for Electric Motors*. London: Chapman and Hall, 1973.

Richardson, Donald V., *Rotating Electric Machinery and Transformer Technology*. Reston, Virginia: Reston Publishing Company, Inc., 1978.

Silicon Controlled Rectifier Designers Handbook, 2nd ed. Leslie R. Rice, ed., Westinghouse Electric Corporation, 1970.

SCR Manual, 5th ed. D. R. Graham and J. C. Hey, eds., General Electric Company, 1972.

SCR Applications Handbook. Dr. Richard G. Hoft, ed. International Rectifier Corp., Semiconductor Division, El Segundo, California, 1974.

Index